Recent Progress in Lithium Niobate

Recent Progress in Lithium Niobate

Editors
Robert A. Jackson
Zsuzsanna Szaller

MDPI • Basel • Beijing • Wuhan • Barcelona • Belgrade • Manchester • Tokyo • Cluj • Tianjin

Editors
Robert A. Jackson
Keele University
UK

Zsuzsanna Szaller
Wigner Research Centre for Physics
Hungary

Editorial Office
MDPI
St. Alban-Anlage 66
4052 Basel, Switzerland

This is a reprint of articles from the Special Issue published online in the open access journal *Crystals* (ISSN 2073-4352) (available at: https://www.mdpi.com/journal/crystals/special_issues/Lithium_Niobate).

For citation purposes, cite each article independently as indicated on the article page online and as indicated below:

LastName, A.A.; LastName, B.B.; LastName, C.C. Article Title. *Journal Name* **Year**, *Article Number*, Page Range.

ISBN 978-3-03943-388-9 (Hbk)
ISBN 978-3-03943-389-6 (PDF)

© 2020 by the authors. Articles in this book are Open Access and distributed under the Creative Commons Attribution (CC BY) license, which allows users to download, copy and build upon published articles, as long as the author and publisher are properly credited, which ensures maximum dissemination and a wider impact of our publications.

The book as a whole is distributed by MDPI under the terms and conditions of the Creative Commons license CC BY-NC-ND.

Contents

About the Editors vii

Preface to "Recent Progress in Lithium Niobate" ix

Robert A. Jackson and Zsuzsanna Szaller
Recent Progress in Lithium Niobate
Reprinted from: *Crystals* 2020, *10*, 780, doi:10.3390/cryst10090780 1

Thomas Kämpfe, Bo Wang, Alexander Haußmann, Long-Qing Chen and Lukas M. Eng
Tunable Non-Volatile Memory by Conductive Ferroelectric Domain Walls in Lithium Niobate Thin Films
Reprinted from: *Crystals* 2020, *10*, 804, doi:10.3390/cryst10090804 5

Romel Menezes Araujo, Emanuel Felipe dos Santos Mattos, Mário Ernesto Giroldo Valerio and Robert A. Jackson
Computer Simulation of the Incorporation of V^{2+}, V^{3+}, V^{4+}, V^{5+} and Mo^{3+}, Mo^{4+}, Mo^{5+}, Mo^{6+} Dopants in $LiNbO_3$
Reprinted from: *Crystals* 2020, *10*, 457, doi:10.3390/cryst10060457 17

Yuejian Jiao, Zhen Shao, Sanbing Li, Xiaojie Wang, Fang Bo, Jingjun Xu and Guoquan Zhang
Improvement on Thermal Stability of Nano-Domains in Lithium Niobate Thin Films
Reprinted from: *Crystals* 2020, *10*, 74, doi:10.3390/cryst10020074 29

Xiaodong Yan, Tian Tian, Menghui Wang, Hui Shen, Ding Zhou, Yan Zhang and Jiayue Xu
High Homogeneity of Magnesium Doped $LiNbO_3$ Crystals Grown by Bridgman Method
Reprinted from: *Crystals* 2020, *10*, 71, doi:10.3390/cryst10020071 39

Huangpu Han, Bingxi Xiang, Tao Lin, Guangyue Chai and Shuangchen Ruan
Design and Optimization of Proton Exchanged Integrated Electro-Optic Modulators in X-Cut Lithium Niobate Thin Film
Reprinted from: *Crystals* 2019, *9*, 549, doi:10.3390/cryst9110549 49

Hongsik Jung
An Integrated Photonic Electric-Field Sensor Utilizing a 1×2 YBB Mach-Zehnder Interferometric Modulator with a Titanium-Diffused Lithium Niobate Waveguide and a Dipole Patch Antenna
Reprinted from: *Crystals* 2019, *9*, 459, doi:10.3390/cryst9090459 57

Oswaldo Sánchez-Dena, Carlos J. Villagómez, César D. Fierro-Ruíz, Artemio S. Padilla-Robles, Rurik Farías, Enrique Vigueras-Santiago, Susana Hernández-López and Jorge-Alejandro Reyes-Esqueda
Determination of the Chemical Composition of Lithium Niobate Powders
Reprinted from: *Crystals* 2019, *9*, 340, doi:10.3390/cryst9070340 69

Laura Kocsor, László Péter, Gábor Corradi, Zsolt Kis, Jenő Gubicza and László Kovács
Mechanochemical Reactions of Lithium Niobate Induced by High-Energy Ball-Milling
Reprinted from: *Crystals* 2019, *9*, 334, doi:10.3390/cryst9070334 87

Liyun Xue, Hongde Liu, Dahuai Zheng, Shahzad Saeed, Xuying Wang, Tian Tian, Ling Zhu, Yongfa Kong, Shiguo Liu, Shaolin Chen, Ling Zhang and Jingjun Xu
The Photorefractive Response of Zn and Mo Codoped $LiNbO_3$ in the Visible Region
Reprinted from: *Crystals* 2019, *9*, 228, doi:10.3390/cryst9050228 103

About the Editors

Robert A. Jackson is Reader in Computational Solid State Chemistry at Keele University. He obtained his BSc, Ph.D. and DSc degrees from University College London, and has published around 150 papers in peer-reviewed journals, including a series of papers on computer modelling of structure, properties and defect chemistry of lithium niobate.

Zsuzsanna Szaller has a Ph.D. degree from Budapest University of Technology and Economics and is a chemical engineer and Ph.D. research fellow in the Institute for Solid State Physics and Optics, Wigner Research Centre for Physics Department of Applied and Nonlinear Optics, Project Crystal Physics. She has 38 published papers. Her research Interests include single crystal growth; niobates; high-temperature top-seeded solution growth.

Preface to "Recent Progress in Lithium Niobate"

LiNbO3 is an all-star material from both scientific and technological viewpoints and has been the subject of a great number of publications since its first preparation in 1937. The journal *Crystals* runs Special Issues to create collections of papers on specific topics. Recent Progress in Lithium Niobate' is the third of these Special Issues dealing with LiNbO3.

Robert A. Jackson, Zsuzsanna Szaller
Editors

Editorial

Recent Progress in Lithium Niobate

Robert A. Jackson [1],* and Zsuzsanna Szaller [2],*

1. Lennard-Jones Laboratories, School of Chemical and Physical Sciences, Keele University, Keele, Staffordshire ST5 5BG, UK
2. Institute for Solid State Physics and Optics, Wigner Research Centre for Physics, Konkoly-Thege Miklós út 29-33, 1121 Budapest, Hungary
* Correspondence: r.a.jackson@keele.ac.uk (R.A.J.); szaller.zsuzsanna@wigner.hu (Z.S.)

Received: 31 August 2020; Accepted: 31 August 2020; Published: 3 September 2020

Abstract: This special issue features eight papers which cover the recent developments in research on lithium niobate. Papers are divided into three groups based on their topic.

Keywords: photorefractive properties; defect structure; lithium niobate

---

1. Photorefractive Properties and Defect Structure

The photorefractive properties of lithium niobate (LN): Mo,Zn crystals with different doping concentrations were investigated in paper [1]. Zinc can shorten the response time and improve the photorefractive sensitivity of the LN:Mo,Zn crystal. Valence states of Mo ions were identified by XPS. Three valences (+6, +5, +4) were identified in the crystal and one (+6) in the residue. In the LN:Mo,Zn 7.2 crystal the $Mo_{Nb}{}^+$ and $Mo_{Li}{}^{3+/4+}$ defects served as the photorefractive centre for fast photorefraction. Potential material for fast response holographic storage are 7.2 mol% Zn and 0.5 mol% Mo co-doped $LiNbO_3$ crystals.

Vanadium and molybdenum ions are of interest in enhancing the photorefractive properties of $LiNbO_3$. Paper [2] presents a computer modelling study of V^{2+}, V^{3+}, V^{4+} and V^{5+} as well as Mo^{3+}, Mo^{4+}, Mo^{5+} and Mo^{6+} in $LiNbO_3$ using interatomic potentials. It was found that divalent (V^{2+}), trivalent (V^{3+}, Mo^{3+}) and tetravalent (V^{4+}) ions are incorporated at the Li and Nb sites through the self-compensation mechanism. However, the tetravalent (Mo^{4+}) ion is more favourably incorporated at the niobium site, compensated by an oxygen vacancy. The pentavalent ions (V^{5+}, Mo^{5+}) and hexavalent (Mo^{6+}) ions substitute Nb. No charge compensation is found for pentavalent ions, but there is charge compensation with a lithium vacancy for the Mo^{6+} ion.

2. $LiNbO_3$ Preparation Techniques

Lithium niobate nanocrystals were prepared by high-energy ball-milling of the residue of a Czochralski grown congruent single crystal which depend on different types of vials, milling parameters as described in paper [3]. Characterisation of LN nanocrystals and mechanochemical reactions of lithium niobate such as decomposition and the redox processes induced by high-energy ball-milling were studied. During the milling process, the formation of the $LiNb_3O_8$ phase taking place and the reaction can be described as

$$3\ LiNbO_3 \rightarrow LiNb_3O_8 + Li_2O \qquad (1)$$

where lithium oxide is a volatile by-product. The material undergoes partial reduction that leads to a balanced formation of bipolarons and polarons yielding a grey colour together with Li_2O segregation on the open surfaces.

In paper [4], determination of chemical composition between congruent and stoichiometric LiNbO$_3$ powders was worked out by four analytical techniques. Sample preparations were done by mechanosynthesis.

In paper [5], Ø2" LN crystals doped with 0.3 mol% and 5 mol% Mg concentrations with high homogeneity were grown by the Bridgman method using a systematically optimised scheme with careful thermal field design. LN:Mg polycrystalline powders were synthesised by a wet chemistry method to avoid scattering particles and inclusions in the crystal. The homogeneity of LN:Mg crystals was also checked. The extraordinary refractive index gradient was as small as 2.5×10^{-5}/cm.

3. Applications of Lithium Niobate Waveguides

Titanium-diffused lithium niobate waveguide devices are suitable for electric-field detection since their sensors will not perturb the field to be measured. Paper [6] studied photonic electric-field sensors using a 1 × 2 Y-fed balanced-bridge Mach-Zehnder interferometer modulator composed of two complementary outputs and a 3 dB directional coupler based on the electro-optic effect and titanium diffused lithium niobate optical waveguides.

Proton-exchange (PE) is one of the waveguide fabricating techniques. In the research in paper [7], authors simulated and analysed a proton-exchanged E-O Mach-Zehnder interferometer in an x-cut lithium niobate on insulator, LNOI. Based on the full-vectorial finite-difference method, the single-mode conditions, mode size, and optical power distribution of PE waveguides were investigated. The bending losses the Y-branch structures were analysed and propagation losses of the PE waveguides with different separation distances between electrodes were simulated. The half-wave voltages of the devices were calculated using the finite difference beam propagation method (FD-BPM).

In paper [8] it was confirmed that the nano-domains in lithium niobate thin films are thermally unstable even at a temperature of the order of ~100 °C, which can be easily reached due to light absorption. The thermal instability of nano-domains could be very detrimental to practical applications, such as periodically poled lithium niobate (PPLN) microcavities, PPLN ridge waveguides, and ferroelectric domain memories. Thermal stability of nano-domains can be greatly improved when the lithium niobate thin film undergoes a pre-heat treatment before the fabrication of nano-domains. This thermal stability improvement is attributed to the generation of a space charge field during the pre-heat treatment, which is parallel to the spontaneous polarisation of nano-domains.

The wide range of topics covered by the papers in this special issue shows that the field of lithium niobate research is very much alive and that we can continue to expect new developments in this research area.

Conflicts of Interest: The authors declare no conflict of interest.

References

1. Xue, L.; Liu, H.; Zheng, D.; Saeed, S.; Wang, X.; Tian, T.; Zhu, L.; Kong, Y.; Liu, S.; Chen, S.; et al. The Photorefractive Response of Zn and Mo Codoped LiNbO$_3$ in the Visible Region. *Crystals* **2019**, *9*, 228. [CrossRef]
2. Araujo, R.; dos Santos Mattos, E.; Valerio, M.; Jackson, R. Computer Simulation of the Incorporation of V^{2+}, V^{3+}, V^{4+}, V^{5+} and Mo^{3+}, Mo^{4+}, Mo^{5+}, Mo^{6+} Dopants in LiNbO$_3$. *Crystals* **2020**, *10*, 457. [CrossRef]
3. Kocsor, L.; Péter, L.; Corradi, G.; Kis, Z.; Gubicza, J.; Kovács, L. Mechanochemical Reactions of Lithium Niobate Induced by High-Energy Ball-Milling. *Crystals* **2019**, *9*, 334. [CrossRef]
4. Sánchez-Dena, O.; Villagómez, C.; Fierro-Ruíz, C.; Padilla-Robles, A.; Farías, R.; Vigueras-Santiago, E.; Hernández-López, S.; Reyes-Esqueda, J. Determination of the Chemical Composition of Lithium Niobate Powders. *Crystals* **2019**, *9*, 340. [CrossRef]
5. Yan, X.; Tian, T.; Wang, M.; Shen, H.; Zhou, D.; Zhang, Y.; Xu, J. High Homogeneity of Magnesium Doped LiNbO$_3$ Crystals Grown by Bridgman Method. *Crystals* **2020**, *10*, 71. [CrossRef]

6. Jung, H. An Integrated Photonic Electric-Field Sensor Utilizing a 1 × 2 YBB Mach-Zehnder Interferometric Modulator with a Titanium-Diffused Lithium Niobate Waveguide and a Dipole Patch Antenna. *Crystals* **2019**, *9*, 459. [CrossRef]
7. Han, H.; Xiang, B.; Lin, T.; Chai, G.; Ruan, S. Design and Optimization of Proton Exchanged Integrated Electro-Optic Modulators in X-Cut Lithium Niobate Thin Film. *Crystals* **2019**, *9*, 549. [CrossRef]
8. Jiao, Y.; Shao, Z.; Li, S.; Wang, X.; Bo, F.; Xu, J.; Zhang, G. Improvement on Thermal Stability of Nano-Domains in Lithium Niobate Thin Films. *Crystals* **2020**, *10*, 74. [CrossRef]

© 2020 by the authors. Licensee MDPI, Basel, Switzerland. This article is an open access article distributed under the terms and conditions of the Creative Commons Attribution (CC BY) license (http://creativecommons.org/licenses/by/4.0/).

Article

Tunable Non-Volatile Memory by Conductive Ferroelectric Domain Walls in Lithium Niobate Thin Films

Thomas Kämpfe [1,2,*], Bo Wang [3], Alexander Haußmann [1], Long-Qing Chen [3] and Lukas M. Eng [1,*]

1. Institute of Applied Physics, TU Dresden, 01069 Dresden, Germany; alexander.haussmann@tu-dresden.de
2. Center Nanoelectronic Technologies, Fraunhofer IPMS, 01099 Dresden, Germany
3. Department of Materials Science and Engineering, and Materials Research Institute, The Pennsylvania State University, University Park, PA 16802, USA; bzw133@psu.edu (B.W.); lqc3@psu.edu (L.-Q.C.)
* Correspondence: thomas.kaempfe@ipms.fraunhofer.de (T.K.); lukas.eng@tu-dresden.de (L.M.E.)

Received: 1 January 2020; Accepted: 26 May 2020; Published: 11 September 2020

Abstract: Ferroelectric domain wall conductance is a rapidly growing field. Thin-film lithium niobate, as in lithium niobate on insulators (LNOI), appears to be an ideal template, which is tuned by the inclination of the domain wall. Thus, the precise tuning of domain wall inclination with the applied voltage can be used in non-volatile memories, which store more than binary information. In this study, we present the realization of this concept for non-volatile memories. We obtain remarkably stable set voltages by the ferroelectric nature of the device as well as a very large increase in the conduction, by at least five orders of magnitude at room temperature. Furthermore, the device conductance can be reproducibly tuned over at least two orders of magnitude. The observed domain wall (DW) conductance tunability by the applied voltage can be correlated with phase-field simulated DW inclination evolution upon poling. Furthermore, evidence for polaron-based conduction is given.

Keywords: conducting domain walls; ferroelectric films; lithium niobate; lithium niobate-on-insulator; scanning probe microscopy; non-volatile memory

---

1. Introduction

In recent years, increasing efforts have been made to develop novel non-volatile memory concepts to meet the increasing demands in terms of scalability and energy consumption. Conductive ferroelectric domain walls (DWs) appear an interesting approach, as ferroelectric DWs are topological defects on the atomic scale and can be created, moved and erased solely by the application of an electric field. Following the discovery of the effect of conductance of DWs in thin-film bismuth ferrite (BFO) [1,2] similar behavior was observed in various other ferroelectric thin films such as lead-zirconate titanate (PZT) [3] and lithium niobate (LNO) [4,5].

The application to non-volatile memories lies in the contradiction in ferroelectrics. Typically, they are known for their very large bandgaps. Thus, one can observe a huge variation in the conductivity between the insulating domain and the conductive domain wall. Various explanations have been made to describe the conductivity of ferroelectric DWs, ranging from oxygen or cation accumulation at the DW to polaron or electron gas formation [6–9] In various reports, the conductivity was proven to be correlated with the charge state of the DW; i.e., DWs inclined to the polar axis showed increased conductance [10–12] Tuning of the DW conductance was also possible by the application of an external field, which resulted in an increase in DW inclination [13].

In this publication, we want to present that precise control over the conductance of domain walls in single-crystalline LNO thin films, thus, implicitly, the inclination angle to the polar axis, can result in an efficient nonvolatile memory element. Moreover, by precise control of the inclination angle, various conductance levels can be distinguished, which is interesting for the application of non-volatile memories. Particularly, multilevel non-volatile memories could be implemented into crossbars to store weight matrixes for neuromorphic computing [14].

2. Materials and Methods

So far, most of the investigations of conductive DWs in LNO have been performed using scanning probe microscopy techniques, such as piezo-response force microscopy (PFM) and conductive-type atomic force microscopy (cAFM), on thick single crystals [15,16] These were backed up by inclination measurements using three-dimensional optical microscopy techniques, such as Cherenkov second-harmonic generation microscopy [17–19] multiphoton microscopy [20–22] and optical coherency microscopy [23], which confirmed the stable inclined DW formation as well as ferroelectric lithography [24]. Furthermore, transmission electron microscopy measurements prove the stable inclination on the atomic scale [25,26].

We investigated single-crystalline congruent thin-film lithium niobate, displaying a single ferroelectric domain after preparation. The samples were fabricated by a modified ion-slicing technique on 6" wafers [27]. Within the process, a platinum electrode is deposited onto the handling wafer before the wafer bonding. The thickness of the layer was set to be 600 nm by chemical-mechanical polishing and checked by ellipsometry. This enables electrical read-out, yet still preserving the single-crystalline and single domain configuration. High-resolution XRD measurements confirmed the high quality of the films oriented along (001). Electrodes with sizes of A ≈ 20,000 μm^2, consisting of Cr/Au with a thickness of 100 nm, were evaporated and lithographically structured to enable the electrical characterization.

The bare film was further investigated by scanning probe microscopy to identify the formation of conductive DWs. Hereby, full-metal Pt AFM tips were applied. The domain patterns were written at a constant voltage of 65 V. To identify the locally written domain pattern we used piezo-response force microscopy (PFM). Conductive AFM (cAFM) was performed at bias voltages up to 10 V.

The local I–V-curves reveal a strong unipolar conductance, which enables erasing the conductive domain walls by applying an external counter-bias. We investigated whether a similar conductive DW formation is possible in a parallel plate capacitor structure using homogeneous electrodes. Hence, we deposited Cr/Au electrodes with a size of A ≈ 20,000 μm^2 and a thickness of 100 nm on a LNO film with a thickness of 600 nm and Ti/Pt back electrode.

3. Results

3.1. Conductive AFM Investigation

DWs were probed by cAFM to investigate the emergence of DW conductivity in these congruent LNO films. In Figure 1a,b a comparison of piezoresponse force microscopy (PFM) and subsequent cAFM measurements is given on the previously created domain pattern, which reveals a perfect match between the derived DWs from PFM and the conductive areas in the films. To further explore the conduction properties of CDWs in LNO thin films, local I–V measurements were carried out. The local current detected for a range of bias voltages is given in Figure 1c,d. Spot 1 is taken as reference and shows only a slight increase in conductivity over as long as about 500s at a voltage of 10 V (<10^{-2} nA). At the DW position, however, there is a strong nonlinear increase in current with applied voltage. At all points, a significant increase over several orders of magnitude (at least three) can be observed, which sufficiently separates these states from the background. The temporal stability measurements show a small increase in current over time. An example is given in Figure 1e. Local probe measurements on these CDWs reveal a stable conductance after 3 h with a minute increase over the measurement of 5% between 1 h and 3 h, hence influences by drift can be ruled out for the given shorter-term measurement.

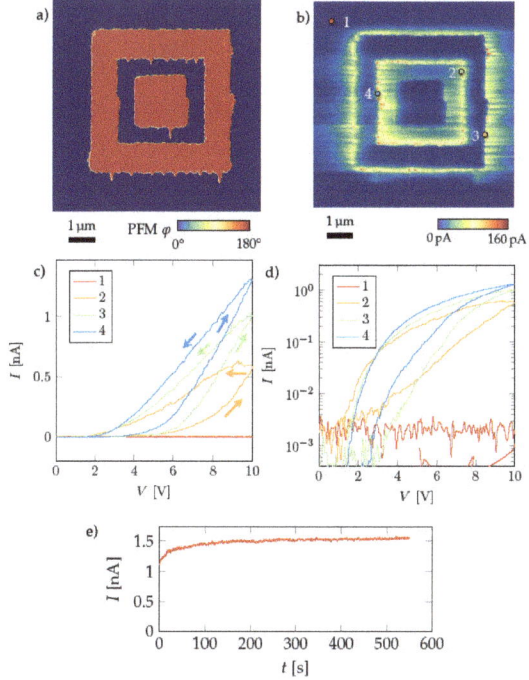

Figure 1. DW conductance in the congruent LNO thin films: (**a**) domain configuration by PFM; (**b**) cAFM scan at a bias voltage of 3 V; (**c**) Local I–V measurements on conductive DWs at four marked spots; (**d**) logarithmic plot of the detected current; (**e**) temporal development of the current at spot 4 at an external bias voltage of 10 V.

3.2. Phase-Field Simulation

The formation of inclined DWs is generally encountered as the reason for conductive DWs in LNO. Yet, the stable formation of such inclined DWs is still a topic of current research. We applied phase field simulations, as they can provide further evidence for the existence and stability of CDWs in ferroelectric thin films. Phase field simulations have been used to study the domain pattern formation in many proper ferroelectrics, including BTO, BFO, and PZT [28]. In this study, we have modeled the temporal evolution of ferroelectric domains in LNO thin films in response to the electrical field created by a biased probe tip. In the simulations, the voltage is ramped up to a given bias voltage. A domain nucleus is formed (shown in Figure 2a), which grows into the single-crystalline film. Due to the external bias field, canted polarization states with in-plane polarization components are created. Afterwards, the inverted domain reaches the rear surface and grows sideways until an equilibrium is reached. When the bias voltage is released, the switched domain relaxes. We observe a stable CDW formation with a non-zero inclination angle, dependent on the maximum bias voltage applied. Since the gradient energy coefficients, which determine the DW width and energy, are rarely reported for LNO, we use the value estimated by Scrymgeour et al. [29]. We notice that increasing the gradient coefficients leads to a disappearance of the DW inclination, suggesting that the stability of inclined DWs in LNO may be attributed to its relatively small gradient energy.

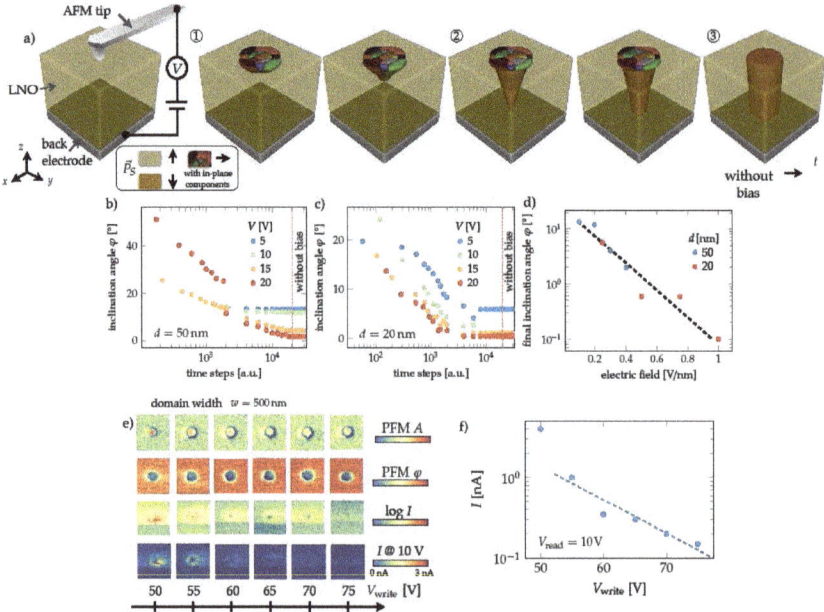

Figure 2. (a) Three-dimensional phase field simulation of CDW formation under an AFM tip for a 20 nm thick lm at 20 V. ① Nucleation, ② through domain formation until equilibrium under external bias, ③ equilibrium domain after removal of bias. (b,c) evolution of inclination angle for film thicknesses, d, of 50 nm and 20 nm respectively; (d) equilibrium inclination angle after removal of bias as a function of applied field; (e) measured current I_{read} at a bias voltage of 10 V for domains written at various writing voltages V_{write}, for comparison PFM scans; (f) extracted maximum domain wall current.

The gradient coefficients of uniaxial LNO and LTO are significantly smaller than for other perovskite ferroelectrics. Previous reports on 180°DW conductance in ferroelectric thin films supports our conjecture. For example, the reported DW inclination in as-poled LTO bulk material was apparently larger [15,30]. Still, it has to be noted that the coefficients are not well known and effects from gradient energy anisotropy and carrier generation could be significant.

The evolution in DW inclination for various tip voltages is given for films of a thickness of 20 nm and 50 nm in Figure 2b,c. In all given cases, a non-zero inclination can be observed after the external bias field is completely removed. The extracted final inclination angle is plotted over the homogenized applied external field. We can observe a decrease in inclination with larger applied external field. A similar behavior in the extracted DW conductance can prove a link between the degree of inclination of a DW and its conductance. To compare the theoretical prediction with the experimental condition, domains were written with a domain size of d = 500 nm at various tip voltages. The written domains are visualized by PFM in Figure 2e. The conductivity extracted by cAFM shows a decrease in current for larger applied writing voltage. In Figure 2f the extracted maximum current values at the DW are given. A similar behavior of the current as a function of the simulated inclination angle can be observed. This is in agreement with previous theoretical assumptions [8] that the DW conductance of inclined DWs is proportional to its inclination and follows $\sigma = 2P_S \sin \alpha$.

3.3. Resistive Switching Investigations

To further analyze the properties of the conductance, I–V measurements were conducted in plate-electrode condition, schematically sketched in Figure 3a under the assumption of inclined domain wall generation at nucleus sites, which would result in strong conductance changes upon reach of the local coercive voltage. Indeed, we can observe a very strong increase over five orders of magnitude in current at a very defined set voltage $V_{set} = 21.05$ V with an accuracy of $\Delta V_{set}/V_{set} = 10^{-3}$, which is proven to be the local coercive voltage from PFM measurements. This value of V_{set} is not only reproduced for a single device, but also for 50 individual devices on a single wafer, which clearly underlines the very precise and reproducible behavior of single-crystalline resistive switching devices. Up to a voltage of -3 V, a very symmetric current–voltage relation can be observed. Yet, for larger negative biases, the absolute value of the current saturates and is not stable anymore but reduces with time. The given cycles in Figure 3d are obtained with a cycle frequency of 1.5 mHz. The observed behavior is very similar to back-switching observed upon current injection from the top electrode at small voltages confirmed by PFM. Hence, we suppose, upon the application of a negative bias, insulating straight or tail-to-tail DWs are formed or domain inversion is invoked; thus, there is no complete conductive channel anymore, which prohibits a current flow.

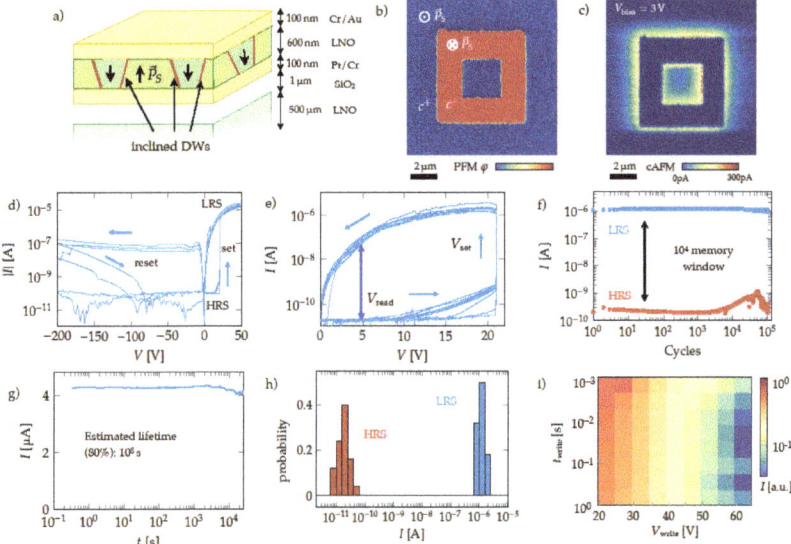

Figure 3. Investigation of the switching behavior, endurance, stability, and tunability of resistive switching of the Pt/LNO/Cr/Au stack with a contact area of 2000 μm². (**a**) film stack configuration (**b**) PFM scan of a written domain (**c**) cAFM scan at a bias voltage of 3V; (**d**) full I-V cycle (f = 1.5 mHz) with a very defined set voltage $V_{set} = 21.05$ V (DVset/Vset~10^{-3}), hence a comparably small electric field $E_{switch,off} = 0.3$ MV/cm and strongly rectifying behavior without significant leakage upon an electric field of $E_{switch,off} = 3.4$ MV/cm with a resistance of >20 TW, (**e**) switch-on I-V cycle with constant switch-off voltage $V_{switch,off} = -210$ V, (**f**) endurance of high resistance and low-resistance state (HRS, LRS, respectively) over at least 10^5 cycles with a resistance window of >10^4 and a read voltage of 10 V, (**g**) time stability of low resistant state over 104 s, which yields an 80% reliability over 108 s or 3 years, (**h**) probability of the current in HRS and LRS at 10 V for 50 tested devices on the same single crystalline thin-film (**i**) tunability of readout current $I_{read,on}$ under modulation of writing time t_{write} and writing voltage V_{write}. The read-out current $I_{read,on}$ reduces for larger writing voltages $V_{write,on}$. $I_{read,on}$ is the average value over 100 writing cycles each.

Before switching the resistance state, one can observe no current larger than 10 pA, which is the lower limit for current detection of the applied source-meter. Further measurements with a further electrometer revealed an even smaller upper current limit of 200 fA at a voltage of −200 V, which corresponds to a resistance of at least 1 PΩ up to a bias voltage of −200 V or an electric field of 3.4 MV/cm, which underlines that leakage is negligible in the films. This similarly holds for positive read-out voltages in the high resistance state (HRS). Hence, the HRS is expected to have a resistance of at least 25 TΩ, with the low resistance state at around 10 MΩ; hence, we observe a huge resistance change over seven orders of magnitude. By setting the switch, no overshoot is present; thus the HRS setting is self-limiting. In Figure 3e the half cycle I–V-curves are shown upon the boundary condition, the current having reduced to 10^{-8} A at −210 V. We can observe the reproducible set voltage. However, below this specific voltage, a slight increase in current can be observed. In general, this current, which deviates from the first cycle, is smaller the lower the current at $V = -210$ V. It saturates after several cycles. Several reasons are possible, e.g., deep traps, which are incorporated into the film upon large current flow. Hence, for endurance testing high voltage treatment was kept as short as possible. In Figure 3f the endurance upon such cycling is shown. The resistance is measured after every cycle. The LRS is created on application of $V_{set} = 21.1$ V. The system is released into the HRS upon application of $V = -210$ V. This results in a very enduring resistive switching device over at least 10^5 cycles. The states can be read out without destruction. The temporal stability of the LRS is given in Figure 3g. As is visible, the conductance is very stable over at least 10^4 s. Hence, assuming an exponential decrease and a minimum current of 80%, a stability over 10^8 s or 10 years can be predicted. Similar measurements with an AFM tip revealed the stability of the current on CDWs over at least 3 h, which is about the longest time to measure due to probe drift. The statistical results given in Figure 3h show a sufficiently high margin for the application as a nonvolatile memory.

Under the application of larger set voltages and shorter pulses being applied among domain formation, the domain wall is expected to straighten, which would result in a decrease in conductance. Indeed, this can be seen as given in Figure 3i. Particularly, the fact that stronger applied electric fields result in reduced conductivity is very unusual and supports the conductance measured to be derived by inclined domain walls.

This proof of high endurance, high on/off current switching, high retention and the tenability among programming makes charged domain walls an interesting tunable nonvolatile memory.

3.4. Conductance Type Extraction

The conduction type was investigated by I–V measurements by linearization with typical mechanisms. In Figure 4a, the I–V measurements are plotted according to space-charge-limited current (SCLC) theory, which predicts I~V^2 in a log–log plot. A fit is given, which shows a perfect fit over almost two voltage magnitudes. In Figure 4b the I–V curve is linearized according to Shottky thermioninc emission (STE). A good fit can be seen in a medium voltage range, but both for a small and large voltage significant deviations are visible. Similarly, in the case of Poole–Frenkel (PF) linearization given in Figure 4c, we obtain a reasonable fit for medium voltages, but discrepancies for very low and high voltages. Yet, it would be not sufficient to formally exclude these two transport mechanisms. Likewise, the linearization with Fowler–Nordheim (FN) tunneling does not give a decent fit. In the LRS we obtain a flat line. Hence, the first three transport mechanisms still have to be considered. Thus, we used the abrupt junction approximation, which assumes a pn-junction with an abrupt doping profile [31] Even though the main advantage of this method is to understand impedance upon a DC bias, it is still helpful to analyze the plain DC I–V curves as dlog(I)/dV is independent on most parameters, especially the contact area; it is, hence, more robust, especially upon temperature changes, and it is possible to differentiate between FN and PF. Hence, in Figure 4d, the I–V curves in a temperature range T between 300 K and 340 K are given. Only SCLC and PF [both dlog(I)/dV ~ V^{-1}] can describe such a curve, whereas STE [dlog(I)/dV ~ $V^{-3/4}$] cannot describe this behavior. Yet, in the case of PF, a very strong component $\propto V^{-1/2}$ should also be present, which cannot be seen in the measurements. The I–V SCLC curve fit is almost perfect over two orders of magnitude and, thus, significantly better than those of PF and STE.

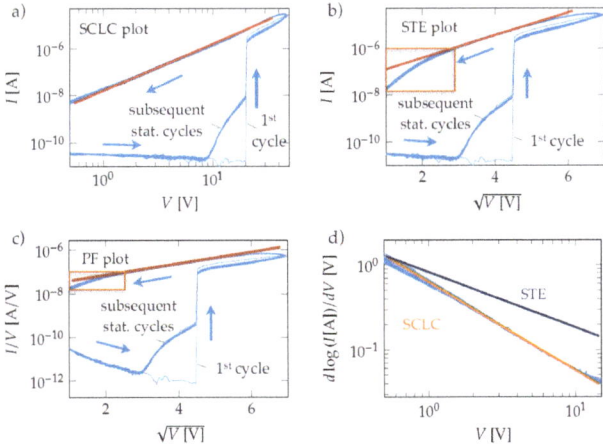

Figure 4. Conductance type analysis on CDWs in LNO (**a**) SCLC I–V plot; (**b**) STE I–V plot; (**c**) PF I–V plot; (**d**) derivative plot. The I–V curves show the first cycle and the stationary cycles after 20 cycles. The off-state was obtained by applying V = −210 V until the current is reduced to 10^{-8} A.

So far, we have solely discussed electronic transport. Yet, especially, resistive switching is generally an interplay of both ionic and electronic current contributions. Hence, it is necessary to further rule out scenarios of mixed electronic–ionic conduction. We use the Nernst–Einstein equation to analyze whether ionic conduction is a significant contribution, which is given by $\Delta = \tau eDE/kT$, with τ the diffusion time and D the diffusion constant. Using a conservative value of D ~ 10^{-18} m^2/s, which is derived from experimental values of lithium transport in Li$_x$Si [32], the estimated ionic movement over 100 ms is about ~0.8 nm at a field of 2×10^7 V/m and room temperature T = 300 K, which is about three orders of magnitude smaller than the film thickness. Reported values for the ionic transport diffusion constants in LNO (e.g., Li, H, D, Na, Mg) at elevated temperatures and interpolated to room temperature are significantly smaller. Hence, we can exclude ionic current having a major share.

The conduction in DWs, thus, follows the Mott–Gurney equation (Child's law) [33]: I(T) = A$_{eff}$ 9εμ(T)V^2/8d^3 with A$_{eff}$ the effective contact area, d the thickness of the dielectric film, ε the static permittivity, and μ(T) the mobility of the major charge carrier. Using the current extracted from cAFM measurements, we can derive a mobility of about $5 \pm 3 \times 10^{-2}$ cm^2/Vs, which is in good agreement with previously reported macroscopic photo-induced current measurements [34]. The temperature dependence in an SCLC transport regime is solely determined by the temperature dependence of the mobility of the major charge carrier.

3.5. Temperature Dependent Conductance

The temperature dependence of the conductance in the Au/Cr/LNO/Pt stack is given in Figure 5. We observe three current regimes. In ①, for temperatures above 300 K, an activation energy of 0.63 eV was obtained; in ②, between 300 K and about 100 K, we obtain an activation energy of 0.18 eV. For temperatures below 100 K ③, we obtain an activation energy of only 0.03 eV. Very similar regimes have been proposed very recently for the lifetime of bound polaron in lightly-doped Fe:LNO via Monte-Carlo simulations [35].

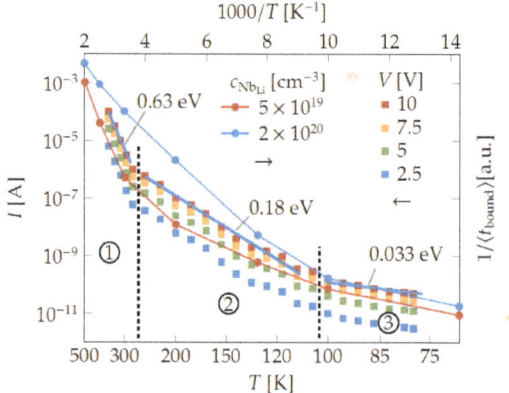

Figure 5. Temperature dependence of the current for thin-film LNO. Measured current at different voltages from 360 K to 77 K. We can observe different regimes of thermal activation ① above 300 K with Ea = 0.63 eV, at ② between 300 K and 100 K with Ea = 0.18 eV and ③ below 100 K with a thermal activation of Ea = 0.03 eV. These are overlaid with the inverse bound polaron lifetime, as calculated by Monte Carlo (MC) simulations.

The processes which are assigned to these three regimes are given as:
① $Nb_{Li}^{4+} + Nb_{Nb}^{5+} \rightarrow Nb_{Li}^{5+} + Nb_{Nb}^{4+}$, Ea ~ 0.65 eV
② $Nb_{Li}^{4+} + Nb_{Li}^{5+} \rightarrow Nb_{Li}^{5+} + Nb_{Li}^{4+}$, Ea ~ 0.2 eV
③ $Nb_{Li}^{4+} + Fe_{Li}^{3+} \rightarrow Nb_{Li}^{5+} + Fe_{Li}^{2+}$, Ea ~ 0.03 eV.

The simulated bound polaron lifetime by Mhaouech and Guilbert is overlaid on the aforementioned temperature-dependent I–V curves given in Figure 5.

Thus, we can conclude that the current through the exfoliated LNO thin film is mainly governed by an interplay of bound and free small polaron transport, depending on the given temperature. As iron impurities act as deep traps, they will inhibit any further conduction. The hopping activation energy of Fe^{2+} is given to be 0.35 eV [36], hence, only a little larger than Li for bound and free polarons, but as the hopping rate follows $w \sim e^{-r/a} \approx e^{-r[\text{Å}]}$, hopping transport is very unlikely for the given concentrations and even for lightly doped Fe:LNO not relevant. A conversion to bound and small polarons is unlikely due to the high binding energy of 1.22 eV.

4. Discussion and Conclusions

In this study, we reported the domain wall conductance by inclined domain walls in LNO thin films, which could be applied as a new multi-level non-volatile memory cell based on out-of-plane ferroelectric switching, by the precise tuning of the inclination and thus the conductance state. We investigated further the temperature dependence as well as the temporal dependence on erase voltage application and concluded that the main contribution to the current is space-charge-limited.

We obtained evidence the current to be dominated by a polaron gas formation at such inclined domain wall and confined by the band bending from the temperature dependence of the conductance. Upon local inversion, generally, charge carrier injection is discussed, even though it has been speculative for a long time. Assuming such charge carriers are injected at the DW, which are needed to compensate the inclined DW of the spike-domain-like nucleus, free charge carriers—electrons—are generated, which will condensate into similarly charged and, hence, screening electron polaronic, states inside the single crystal material. Such a transition has been observed in bulk material, e.g., by optical time-resolved spectroscopy, to happen within 200 fs [37]. Yet, at a CDW these values can be significantly larger due to the smaller energy gap between conductance band and Fermi energy. However, as the band-bending is not large enough to create a 2DEG, which would be apparent with much higher mobility conductivity, we assume a 2D polaron gas (2DPG) is obtained. Bound polarons will very rapidly be trapped by iron impurities. Upon pure-diffusion transport, thus without an external electric field, the average mean free path of the bound polaron until trapping with Fe_{Li} is < $t_{NbLi4+} \geq 100$ Å and for the free polaron is < $t_{NbNb4+} \geq 80$ Å. These values were derived by interpolation from the data of Mhaouech and Guilbert [35] under consideration of the smaller iron impurity concentration $c_{Fe} = 10^{17}$ cm^{-3} and a niobium antisite defect concentration of $c_{NbLi} = 10^{21}$ cm^{-3}. Upon drift condition, these values can be larger. Hence, the bound polaron has a larger mean free path. However, the hopping rate of bound polarons is only $w_{NbLi4+} = 10^7$ s^{-1}. Assuming the hopping rate to depend mainly on the nearest neighbor distance, like $w \sim e^{r[Å]}$, we can calculate, that the hopping rate of free polarons is two orders of magnitude larger, hence $w_{NbLi4+} = 10^9$ s^{-1}. Thus, the electric current with free polaron transport as in Mg:LNO can be significantly larger. This again explains why, for the case of bulk material only, congruent Mg:LNO revealed a conductance upon UV illumination. For the ultrathin LNO films in use in this study, even undoped material can give measureable conductance.

The model explains why an asymmetric electrode condition can result in the efficient switching between an HRS and LRS. Upon the negative pulse application, the carrier density reduces, which destabilizes the inclined domain wall formation and finally reduces the conductance of the wall further. However, this process requires relatively high electric fields and times due to the low mobility of the charge carriers. Hence, the study suggests that Mg:LNO and thinner films can significantly improve the performance of the tunable non-volatile memory.

The thin-film lithium niobate resistive switching effect based on charged ferroelectric domain walls formation and erasure offers interesting advantages over conventional filament-based resistive switching, which can be interesting for neuromorphic computing applications.

Author Contributions: Investigation, T.K., B.W.; resources, A.H., L.M.E.; writing—review and editing, T.K.; supervision, L.-Q.C., L.M.E. All authors have read and agreed to the published version of the manuscript.

Funding: This research was funded by the Cluster of Excellence "Center of Advancing Electronics Dresden" and the DFG Research Grant HA 6982/1-1. We also acknowledge open access funding support by the publication fund of TU Dresden.

Acknowledgments: The authors would like to acknowledge the support of NanoLN for the provision of the lithium niobate thin film samples, H. Hui (Shandong University) for helpful discussions, and S. Johnston as well as Z.X. Shen (Stanford) for measurement support. B.W. acknowledges the support by the NSF-MRSEC grant number DMR-1420620. The effort of L.-Q.C. is supported by National Science Foundation (NSF) through Grant No. DMR-1744213.

Conflicts of Interest: The authors declare no conflict of interest.

References

1. Seidel, J.; Martin, L.W.; He, Q.; Zhan, Q.; Chu, Y.-H.; Rother, A.; Hawkridge, M.E.; Maksymovych, P.; Yu, P.; Gajek, M.; et al. Conduction at domain walls in oxide multiferroics. *Nat. Mater.* **2009**, *8*, 229–234. [CrossRef] [PubMed]
2. Guyonnet, J.; Gaponenko, I.; Gariglio, S.; Paruch, P. Conduction at domain walls in insulating Pb ($Zr_{0.2}Ti_{0.8}$)O_3 thin films. *Adv. Mater.* **2011**, *23*, 5377. [CrossRef] [PubMed]
3. Gaponenko, I.; Tückmantel, P.; Karthik, J.; Martin, L.W.; Paruch, P. Towards reversible control of domain wall conduction in Pb ($Zr_{0.2}Ti_{0.8}$)O_3 thin films. *Appl. Phys. Lett.* **2015**, *106*, 162902. [CrossRef]
4. Lu, H.; Tan, Y.; McConville, J.P.V.; Ahmadi, Z.; Wang, B.; Conroy, M.; Moore, K.; Bangert, U.; Shield, J.E.; Chen, L.-Q.; et al. Electrical Tunability of Domain Wall Conductivity in LiNbO$_3$ Thin Films. *Adv. Mater.* **2019**, *31*, 1902890. [CrossRef]
5. Volk, T.R.; Gainutdinov, R.V.; Zhang, H.H. Domain-wall conduction in AFM-written domain patterns in ion-sliced LiNbO$_3$ films. *Appl. Phys. Lett.* **2017**, *110*, 132905. [CrossRef]
6. Rojac, T.; Bencan, A.; Drazic, G.; Sakamoto, N.; Ursic, H.; Jancar, B.; Tavcar, G.; Makarovic, M.; Walker, J.; Malic, B.; et al. Domain-wall conduction in ferroelectric BiFeO$_3$ controlled by accumulation of charged defects. *Nat. Mater.* **2017**, *16*, 322–327. [CrossRef]
7. Eliseev, E.A.; Morozovska, A.N.; Svechnikov, G.S.; Gopalan, V.; Shur, V.Y. Static conductivity of charged domain walls in uniaxial ferroelectric semiconductors. *Phys. Rev. B* **2011**, *83*, 235313. [CrossRef]

8. Xiao, S.Y.; Kämpfe, T.; Jin, Y.M.; Haußmann, A.; Lu, X.M.; Eng, L.M. Dipole-Tunneling Model from Asymmetric Domain-Wall Conductivity in LiNbO$_3$ Single Crystals. *Phys. Rev. Appl.* **2018**, *10*, 034002. [CrossRef]
9. Sturman, B.; Podivilov, E.; Stepanov, M.; Tagantsev, A.; Setter, N. Quantum properties of charged ferroelectric domain walls. *Phys. Rev. B* **2015**, *92*, 21. [CrossRef]
10. Godau, C.; Kämpfe, T.; Thiessen, A.; Eng, L.M.; Haußmann, A. Enhancing the domain wall conductivity in lithium niobate single crystals. *ACS Nano* **2017**, *11*, 5.
11. Werner, C.S.; Herr, S.J.; Buse, K.; Sturman, B.; Soergel, E.; Razzaghi, C.; Breunig, I. Large and accessible conductivity of charged domain walls in lithium niobate. *Sci. Rep.* **2017**, *7*, 9862. [CrossRef] [PubMed]
12. Pawlik, A.-S.; Kämpfe, T.; Haußmann, A.; Woike, T.; Treske, U.; Knupfer, M.; Büchner, B.; Soergel, E.; Streubel, R.; Koitzsch, A.; et al. Polarization driven conductance variations at charged ferroelectric domain walls. *Nanoscale* **2017**, *9*, 30. [CrossRef] [PubMed]
13. Kirbus, B.; Godau, C.; Wehmeier, L.; Beccard, H.; Beyreuther, E.; Haußmann, A.; Eng, L.M. Real-Time 3D Imaging of Nanoscale Ferroelectric Domain Wall Dynamics in Lithium Niobate Single Crystals under Electric Stimuli: Implications for Domain-Wall-Based Nanoelectronic Devices. *ACS Appl. Nanomater.* **2019**, *2*, 5787–5794. [CrossRef]
14. Xia, Q.; Yang, J.J. Memristive crossbar arrays for brain-inspired computing. *Nat. Mater.* **2019**, *18*, 309–323. [CrossRef] [PubMed]
15. Schröder, M.; Haußmann, A.; Thiessen, A.; Soergel, E.; Woike, T.; Eng, L.M. Conducting domain walls in lithium niobate single crystals. *Adv. Funct. Mater.* **2012**, *22*, 3936. [CrossRef]
16. Schröder, M.; Chen, X.; Haußmann, A.; Thiessen, A.; Poppe, J.; Bonnell, D.A.; Eng, L.M. Nanoscale and macroscopic electrical ac transport along conductive domain walls in lithium niobate single crystals. *Mater. Res. Express* **2014**, *1*, 035012. [CrossRef]
17. Kämpfe, T.; Reichenbach, P.; Schröder, M.; Haußmann, A.; Eng, L.M.; Woike, T.; Soergel, E. Optical three-dimensional profiling of charged domain walls in ferroelectrics by Cherenkov second-harmonic generation. *Phys. Rev. B* **2014**, *89*, 035314. [CrossRef]
18. Kämpfe, T.; Reichenbach, P.; Haußmann, A.; Woike, T.; Soergel, E.; Eng, L.M. Real-time three-dimensional profiling of ferroelectric domain walls. *Appl. Phys. Lett.* **2015**, *107*, 152905. [CrossRef]
19. Wehmeier, L.; Kämpfe, T.; Haußmann, A.; Eng, L.M. In situ 3D observation of the domain wall dynamics in a triglycine sulfate single crystal upon ferroelectric phase transitio. *Phys. Stat. Solidi RRL* **2017**, *11*, 1700267. [CrossRef]
20. Reichenbach, P.; Kämpfe, T.; Thiessen, A.; Haußmann, A.; Woike, T.; Eng, L.M. Multiphoton photoluminescence contrast in switched Mg: LiNbO$_3$ and Mg: LiTaO$_3$ single crystals. *Appl. Phys. Lett.* **2014**, *105*, 22906. [CrossRef]
21. Reichenbach, P.; Kämpfe, T.; Thiessen, A.; Schröder, M.; Haußmann, A.; Woike, T.; Eng, L.M. Conducting domain walls in lithium niobate single crystals. *J. Appl. Phys.* **2014**, *115*, 213509. [CrossRef]
22. Reichenbach, P.; Kämpfe, T.; Haußmann, A.; Thiessen, A.; Woike, T.; Steudtner, R.; Kocsor, L.; Szaller, Z.; Kovács, L.; Eng, L.M. Polaron-Mediated Luminescence in Lithium Niobate and Lithium Tantalate and Its Domain Contrast. *Crystals* **2018**, *8*, 214. [CrossRef]
23. Haußmann, A.; Kirsten, L.; Schmidt, S.; Cimalla, P.; Wehmeier, L.; Koch, E.; Eng, L.M. Three-Dimensional, Time-Resolved Profiling of Ferroelectric Domain Wall Dynamics by Spectral-Domain Optical Coherence Tomography. *Ann. Phys.* **2017**, *529*, 1700139. [CrossRef]
24. Haußmann, A.; Gemeinhardt, A.; Schröder, M.; Kämpfe, T.; Eng, L.M. Bottom-up assembly of molecular nanostructures by means of ferroelectric lithography. *Langmuir* **2017**, *33*, 475–484. [CrossRef]
25. Gonnissen, J.; Batuk, D.; Nataf, G.F.; Jones, L.; Abakumov, A.M.; van Aert, S.; Schryvers, D.; Salje, E.K.H. Direct Observation of Ferroelectric Domain Walls in LiNbO$_3$: Wall-Meanders, Kinks, and Local Electric Charges. *Adv. Funct. Mater.* **2016**, *26*, 42. [CrossRef]
26. Conroy, M.; Moore, K.; O'Connell, E.N.; McConville, J.P.V.; Lu, H.; Chaudhary, P.; Lipatov, A.; Sinitskii, A.; Gruverman, A.; Gregg, J.M.; et al. Atomic-Scale Characterization of Ferro-Electric Domains in Lithium Niobate-revealing the Electronic Properties of Domain Wall. *Microsc. Microanal.* **2019**, *25*, 576–577. [CrossRef]
27. Poberaj, G.; Hu, H.; Sohler, W.; Günter, P. Lithium niobate on insulator (LNOI) for micro-photonic devices. *Laser Photonics Rev.* **2012**, *6*, 488–503. [CrossRef]
28. Chen, L.Q. Phase-field models for microstructure evolution. *Ann. Rev. Mat. Res.* **2002**, *32*, 113–140. [CrossRef]
29. Scrymgeour, D.; Gopalan, V.; Itagi, A.; Saxena, A.; Swart, P. Phenomenological theory of a single domain wall in uniaxial trigonal ferroelectrics: Lithium niobate and lithium tantalite. *Phys. Rev. B* **2005**, *71*, 184110. [CrossRef]

30. Kämpfe, T. Charged Domain Walls in Ferroelectric Single Crystals. Ph.D. Thesis, Dresden University of Technology, Dresden, Germany, 2017.
31. Maksymovych, P.; Pan, M.; Yu, P.; Ramesh, R.; Baddorf, A.P.; Kalinin, S.V. Scaling and disorder analysis of local I–V curves from ferroelectric thin films of lead zirconate titanate. *Nanotechnology* **2011**, *22*, 254031. [CrossRef]
32. Ding, N.; Xu, J.; Yao, Y.X.; Wegner, G.; Fang, X.; Chen, C.H.; Lieberwirth, I. Determination of the diffusion coefficient of lithium ions in nano-Si. *Solid State Ion.* **2009**, *180*, 222–225. [CrossRef]
33. Mott, N.F.; Gurney, R.W. *Electronic Processes in Ionic Crystals*; The Clarendon Press: Oxford, UK, 1940.
34. Kämpfe, T.; Haußmann, A.; Eng, L.M.; Reichenbach, P.; Thiessen, A.; Woike, T.; Steudtner, R. Time-resolved photoluminescence spectroscopy of Nb Nb 4+ and O− polarons in $LiNbO_3$ single crystals. *Phys. Rev. B* **2016**, *93*, 74116. [CrossRef]
35. Mhaouech, I.; Guilbert, L. Temperature dependence of small polaron population decays in iron-doped lithium niobate by Monte Carlo simulations. *Solid State Sci.* **2016**, *60*, 28–36. [CrossRef]
36. Sanson, A.; Zaltron, A.; Argiolas, N.; Sada, C.; Bazzan, M.; Schmidt, W.G.; Sanna, S. Polaronic deformation at the $Fe^{2+/3+}$ impurity site in Fe: $LiNbO_3$ crystals. *Phys. Rev. B* **2015**, *91*, 094109. [CrossRef]
37. Schirmer, O.F.; Imlau, M.; Merschjann, C.; Schoke, B. Electron small polarons and bipolarons in $LiNbO_3$. *J. Phys. Cond. Matter* **2009**, *21*, 23201. [CrossRef]

© 2020 by the authors. Licensee MDPI, Basel, Switzerland. This article is an open access article distributed under the terms and conditions of the Creative Commons Attribution (CC BY) license (http://creativecommons.org/licenses/by/4.0/).

Article

Computer Simulation of the Incorporation of V^{2+}, V^{3+}, V^{4+}, V^{5+} and Mo^{3+}, Mo^{4+}, Mo^{5+}, Mo^{6+} Dopants in LiNbO$_3$

Romel Menezes Araujo [1,2], Emanuel Felipe dos Santos Mattos [1], Mário Ernesto Giroldo Valerio [3] and Robert A. Jackson [4,*]

1. Chemistry Coordination/IPISE/PIC, Pio Decimo College, Campus Jabotiana, Aracaju-SE 49027-210, Brazil; raraujoster@gmail.com (R.M.A.); manel12309@gmail.com (E.F.d.S.M.)
2. Research Institute—Instituto de Pesquisa, Tecnologia e Negócios-IPTN, Aracaju-SE 49095000, Brazil
3. Physics Department, Federal University of Sergipe, Campus Universitário, São Cristovão-SE 491000-000, Brazil; megvalerio@gmail.com
4. Lennard-Jones Laboratories, School of Chemical and Physical Sciences, Keele University, Keele, Staffordshire ST5 5BG, UK
* Correspondence: r.a.jackson@keele.ac.uk

Received: 10 March 2020; Accepted: 6 May 2020; Published: 1 June 2020

Abstract: The doping of LiNbO$_3$ with V^{2+}, V^{3+}, V^{4+} and V^{5+} as well as Mo^{3+}, Mo^{4+}, Mo^{5+} and Mo^{6+} ions is of interest in enhancing its photorefractive properties. In this paper, possible incorporation mechanisms for these ions in LiNbO$_3$ are modelled, using a new set of interaction potentials fitted to the oxides VO, V$_2$O$_3$, VO$_2$, V$_2$O$_5$ and to LiMoO$_2$, Li$_2$MoO$_3$, LiMoO$_3$, Li$_2$MoO$_4$.

Keywords: lithium niobate; divalent; trivalent; tetravalent; pentavalent and hexavalent doping; computer modelling

1. Introduction

Ferroelectric lithium niobate is a material that has been extensively studied because of its many technological applications, including optical integrated circuits, electro-optical modulators, optical memories, acoustic filters, high-frequency beam deflectors, frequency converters and holographic volume storage [1–9], for which holographic volume storage performance is very important [10–15]. This paper looks at the doping of LiNbO$_3$ with vanadium and molybdenum ions in different charge states, with the aim of predicting the optimum location of dopants, and charge compensation mechanisms where needed.

Previous work on vanadium and molybdenum doped lithium niobate has included experimental studies of how its photorefractive properties are enhanced by doping with molybdenum ions [16,17] where it is suggested that the Mo^{6+} ion dopes at the Nb^{5+} site. Another study looks at LiNbO$_3$ co-doped with Mg and V, concluding that some of the vanadium dopes at the Nb site in the 5+ charge state, but that $V^{4+}{}_{Li}$, $V^{3+}{}_{Li}$ and $V^{2+}{}_{Li}$ defects are also observed [18]. Finally, another recent publication [19] has looked at the photorefractive response of Zn and Mo co-doped LiNbO$_3$ in the visible region, and concluded that the presence of Mo^{6+} ions helps promote fast response and multi-wavelength holographic storage, which is attributed to their occupation of regular niobium sites in the lattice.

In a Density Functional Theory (DFT) study [20], vanadium doping was modelled, and it was concluded that vanadium substitutes at the Li$^+$ site as V^{4+}, but that it dopes at the Nb site as a neutral defect as the Fermi level is increased. In another DFT study [21], molybdenum doping was modelled and it was concluded that the most stable configuration involves doping at the Nb^{5+} site, in agreement with the previously mentioned experimental studies [16,17]. It is noted that in the DFT studies, predictions were made on the basis of defect formation energies, as opposed to the solution energy approach adopted in this paper.

This paper presents a computer modelling study of V^{2+}, V^{3+}, V^{4+} and V^{5+} as well as Mo^{3+}, Mo^{4+}, Mo^{5+} and Mo^{6+} doping in $LiNbO_3$ using interatomic potentials. Such calculations enable predictions to be made of the sites occupied by dopant ions, and the form of charge compensation adopted, if needed. These calculations provide information about how the defects behave in the material, and how they influence its properties in the applications mentioned previously. It follows a series of papers by the authors on $LiNbO_3$ doped with a range of ions [22–27].

2. Materials and Methods

2.1. Interatomic Potentials

The interatomic potentials used in this work consist of Buckingham potentials, supplemented by an electrostatic term, as given below:

$$V(r_{ij}) = \frac{q_i q_j}{r_{ij}} + A_{ij} \exp\left(\frac{-r_{ij}}{\rho_{ij}}\right) - C_{ij} r_{ij}^{-6} \tag{1}$$

This expression shows that for each pair of ions it is necessary to determine three parameters: A_{ij}, ρ_{ij} and C_{ij}, which are constants for each interaction, q_i, q_j represent the charges of the ions i and j, and r_{ij} is the interatomic distance. The parameters are determined by empirical fitting, and formal charges are used for q_i and q_j. The procedure by which potentials were obtained for $LiNbO_3$ is explained in the work of Jackson and Valério [22], and the derivation of the potentials for the vanadium and molybdenum dopants is described in Section 3.1 below. The potentials for $LiNbO_3$ have been the subject of recent studies on the doping of the structure with rare earth ions [23,24], doping with Sc, Cr, Fe and In [25], metal co-doping [26] and doping with Hf [27]. These papers show that modelling can predict the energetically optimal locations of the dopant ions and calculate the energy involved in the doping process. This paper extends this procedure to the study of V^{2+}, V^{3+}, V^{4+} and V^{5+} as well as Mo^{3+}, Mo^{4+}, Mo^{5+} and Mo^{6+} doped lithium niobate, with the aim of establishing the optimal doping site and charge compensation scheme for both sets of ions.

2.2. Defect Formation Energies

The calculation of defect formation energies is carried out using the Mott–Littleton approximation [28], in which the crystal is divided into two regions: region I, which contains the defect, and region II, which extends from the edge of region I to infinity. In region I, the positions of the ions are adjusted until the resulting force is zero. The radius of region I is selected such that the forces in region II are relatively weak and the relaxation can be treated according to the harmonic response to the defect (a dielectric continuum). An interfacial region IIa is introduced to treat interactions between region I and region II.

3. Results and Discussion

3.1. Derivation of Interatomic Potential Parameters

It was necessary to derive potential parameters for the dopant oxide structures: VO, V_2O_3, VO_2 and V_2O_5 as well as $LiMoO_2$, Li_2MoO_3, Li_3MoO_4 and Li_2MoO_4. For V^{2+}-O^{2-}, V^{3+}-O^{2-}, V^{4+}-O^{2-} and V^{5+}-O^{2-} as well as Mo^{3+}-Li^+, Mo^{4+}-Li^+, Mo^{5+}-Li^+, Mo^{6+}-Li^+, Mo^{3+}-O^{2-}, Mo^{4+}-O^{2-}, Mo^{5+}-O^{2-} and Mo^{6+}-O^{2-} interactions, a new set of potentials was derived empirically by fitting to the observed structures as shown in Table 1. The O^{2-}-O^{2-} potential was obtained by Sanders et al. [29] and uses the shell model for O [30], which is a representation of ionic polarisability, in which each ion is represented by a core and a shell, coupled by a harmonic spring, and the Li-O potential was taken from [22]. In all cases, the dopant-oxide potentials were obtained by fitting to parent oxide structures.

Table 1. Interionic potentials obtained from a fit to the VO, V_2O_3, VO_2, V_2O_5, $LiMoO_2$, Li_2MoO_3, Li_3MoO_4 and Li_2MoO_4 structures.

Interaction	A_{ij}(eV)	ρ_{ij}(Å)	C_{ij}(Å^6 eV)
Li_{core}-O_{shell}	950.0	0.2610	0.0
V_{core}-O_{shell}	293.240087	0.475181	0.0
Mo_{core}-Li_{core}	573.532325	0.369602	0.0
Mo_{core}-O^{2-}_{shell}	3003.79	0.3474	0.0
Mo_{core}-O_{core}	600.263736	0.328558	0.0
O^{2-}_{shell}-O^{2-}_{shell}	22764.0	0.1490	27.88
Harmonic	k(eV Å^2)	r_0(Å)	
V_{core}-O_{core}	46.997833	1.942956	
Mo_{core}-O_{core}	385.638986	2.073074	
Species			Y(e)
Mo_{core}			3.0 4.0 5.0 6.0
V_{core}			2.0 3.0 4.0 5.0
O_{core}			0.9
O_{shell}			−2.9
Spring			k(Å^{-2} eV)
O_{core}-O_{oore}			70.0

Table 2 compares experimental and calculated structures of VO [31], V_2O_3 [32], VO_2 [33] and V_2O_5 [34] oxides as well as $LiMoO_2$ [35], Li_2MoO_3 [36], Li_3MoO_4 [37] and Li_2MoO_4 [38] lithium molybdate structures, using the potentials in Table 1. It is seen that the experimental and calculated lattice parameters differ by less than 1%, confirming that the potentials can be used in further simulations of defect properties. The calculations were carried at 0 K (the default for the modelling code and used in most other theoretical studies) and at 293 K for comparison with room temperature results. In this way, we can see how the structure and energies vary with temperature.

Table 2. Comparison of calculated (calc.) and experimental (expt.) lattice parameters.

Oxide	Lattice Parameter	Exp.	Calc. (0 K)	Δ%	Calc. (293 K)	Δ%
VO	a(Å) = b(Å) = c(Å)	4.067800	4.108237	0.99	4.10683	0.98
V_2O_3	a(Å) = b(Å) = c(Å)	9.393000	9.304757	0.90	9.346331	0.94
VO_2	a (Å) = b(Å)	4.556100	4.569483	0.20	4.566212	0.22
	c(Å)	2.859800	2.866421	0.23	2.857861	0.07
	a(Å)	11.971900	11.99652	0.20	12.01247	0.33
V_2O_5	b(Å)	4.701700	4.722561	0.44	4.660343	0.88
	c(Å)	5.325300	5.355671	0.57	5.371149	0.86
Lithium Molybdates	Lattice Parameter	Exp.	Calc. (0 K)	Δ%	Calc. (293 K)	Δ%
$LiMoO_2$	a(Å) = b(Å)	2.866300	2.880528	0.50	2.887246	0.73
	c(Å)	15.474300	15.409390	0.42	15.595024	0.78
Li_2MoO_3	a(Å) = b(Å)	2.878000	2.854443	0.82	2.859809	0.63
	c(Å)	14.91190	15.002886	0.61	15.04632	0.90
Li_3MoU_4	a(Å) = b(Å) = c(Å)	4.1389	4.107762	0.75	4.106941	0.77
Li_2MoO_4	a(Å) = b(Å)	14.330000	14.301305	0.20	14.384501	0.38
	c(Å)	9.584	9.492067	0.96	9.632413	0.96

3.2. Defect Calculations

In this section, calculated energies for dopant ions in $LiNbO_3$ are reported. The divalent, trivalent, tetravalent, pentavalent and hexavalent dopants can substitute at Li and Nb sites in the $LiNbO_3$ matrix with charge compensation taking place in a number of ways. The proposed schemes described in the following subsections are written as solid state reactions using the Kroger–Vink notation [39]. This notation appears in the tables in Sections 3.2.1–3.2.5 where the dot/bullet (·) means a net positive charge and the dash/prime (′) means a net negative charge.

3.2.1. Divalent Dopants

The substitution of the divalent dopant V^{2+} in the Li^+ and Nb^{5+} host sites requires a charge-compensating defect, which can involve Li and Nb vacancies, Nb_{Li} anti-sites, interstitial oxygen, self-compensation and oxygen vacancies. The modes of substitution considered for divalent cations are shown in Table 3.

Table 3. Types of defects considered due to M = V^{2+} incorporation in LiNbO$_3$.

Site	Charge Compensation	Reaction
Li^+	Lithium Vacancies	(i) $MO + 2 Li_{Li} \rightarrow M_{Li}^{\cdot} + V'_{Li} + Li_2O$
	Niobium Vacancies	(ii) $5MO + 5Li_{Li} + Nb_{Nb} \rightarrow 5M_{Li}^{\cdot} + V'''''_{Nb} + 2.5Li_2O + 0.5Nb_2O_5$
	Oxygen Interstitial	(iii) $2MO + 2Li_{Li} \rightarrow 2M_{Li}^{\cdot} + O''_i + Li_2O$
Li^+ and Nb^{5+}	Self-Compensation	(iv) $4MO + 3 Li_{Li} + Nb_{Nb} \rightarrow 3M_{Li}^{\cdot} + M'''_{Nb} + 1.5Li_2O + 0.5 Nb_2O_5$
Nb^{5+}	Lithium Vacancies and Anti-site (Nb_{Li})	(v) $MO + 2Li_{Li} + Nb_{Nb} \rightarrow M'''_{Nb} + V'_{Li} + Nb_{Li} + Li_2O$
	Anti-site (Nb_{Li})	(vi) $4MO + 3Li_{Li} + 4Nb_{Nb} \rightarrow 4M'''_{Nb} + 3Nb_{Li} + Li_2O + LiNbO_3$
		(vii) $4MO + 3Li_{Li} + 4Nb_{Nb} \rightarrow 4M'''_{Nb} + 3Nb_{Li} + 1.5Li_2O + 0.5Nb_2O_5$
	Oxygen Vacancies	(viii) $2MO + 2Nb_{Nb} + 3O_O \rightarrow 2M'''_{Nb} + 3V_O^{\cdot\cdot} + Nb_2O_5$

The solution energies for the divalent (V^{2+}) dopant with different charge-compensating mechanisms were evaluated and plotted as a function of the reaction schemes. Based on the lowest energy value, it seems that the incorporation of a divalent (V^{2+}) ion is energetically favourable at the lithium and niobium sites, taking into account the first in relation to the c axis. In schemes (i) and (iv), the energy difference in eV is small at both temperatures in the first neighbours, indicating that it can be incorporated at the lithium site compensated by a lithium vacancy as well as by self-compensation as shown in Figure 1. This can be attributed to the similarity between the ionic radius of V^{2+}, which is 0.79 Å, and those of the Li^+ site, which varies between 0.59 and 0.74 Å, and the Nb^{5+} site, which varies between 0.32 and 0.71 Å [40].

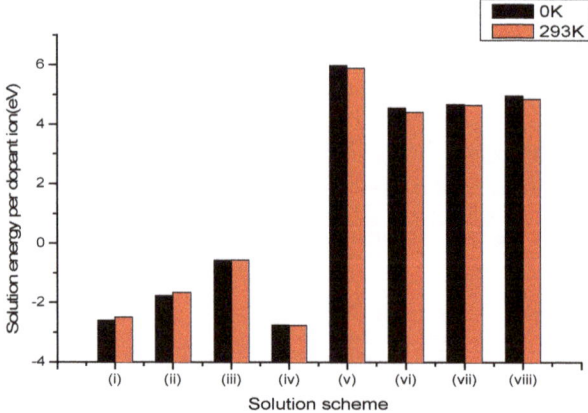

Figure 1. Bar chart of solution energies vs. solution schemes for divalent dopant (V^{2+}) at the Li and Nb sites, considering the first neighbours in relation to the c axis.

3.2.2. Trivalent Dopants

As with the divalent ion V^{2+}, the trivalent V^{3+} and Mo^{3+} dopants can be incorporated at the lithium and niobium sites in the $LiNbO_3$ matrix through various schemes as shown in Tables 4 and 5. When these ions are substituted at Li and Nb sites, the extra positive charge can, as noted earlier, be compensated by the creation of vacancies, interstitials, anti-site defects or self-compensation.

Table 4. Types of defects considered due to V^{3+} incorporation in $LiNbO_3$.

Site	Charge Compensation	Reaction
Li^+	Lithium Vacancies	(i) $0.5M_2O_3 + 3Li_{Li} \rightarrow M_{Li}^{\cdot\cdot} + 2V'_{Li} + 1.5Li_2O$
	Niobium Vacancies	(ii) $2.5M_2O_3 + 5Li_{Li} + 2Nb_{Nb} \rightarrow 5M_{Li}^{\cdot\cdot} + 2V''''_{Nb} + 2.5Li_2O + Nb_2O_5$
	Oxygen Interstitial	(iii) $0.5M_2O_3 + Li_{Li} \rightarrow M_{Li}^{\cdot\cdot} + O''_i + 0.5Li_2O$
Li^+ and Nb^{5+}	Self-Compensation	(iv) $M_2O_3 + Li_{Li} + Nb_{Nb} \rightarrow M_{Li}^{\cdot\cdot} + M''_{Nb} + 0.5Li_2O + 0.5Nb_2O_5$
Nb^{5+}	Oxygen Vacancies	(v) $0.5M_2O_3 + Nb_{Nb} + O_O \rightarrow M''_{Nb} + V_O^{\cdot\cdot} + 0.5Nb_2O_5$
	Anti-site (Nb_{Li})	(vi) $M_2O_3 + Li_{Li} + 2Nb_{Nb} \rightarrow 2M''_{Nb} + Nb_{Li}^{\cdot\cdot\cdot\cdot} + LiNbO_3$
	Lithium Vacancies and Anti-site (Nb_{Li})	(vii) $0.5M_2O_3 + 3Li_{Li} + Nb_{Nb} \rightarrow M''_{Nb} + 2V'_{Li} + Nb_{Li}^{\cdot\cdot\cdot\cdot} + 1.5Li_2O$

Table 5. Types of defects considered due to Mo^{3+} incorporation in $LiNbO_3$.

Site	Charge Compensation	Reaction
Li^+	Lithium Vacancies	(i) $LiMoO_2 + 3Li_{Li} \rightarrow Mo_{Li}^{\cdot\cdot} + 2V'_{Li} + 2Li_2O$
	Niobium Vacancies	(ii) $5LiMoO_2 + 5Li_{Li} + 2Nb_{Nb} \rightarrow 5Mo_{Li}^{\cdot\cdot} + 2V''''_{Nb} + 5Li_2O + Nb_2O_5$
	Oxygen Interstitial	(iii) $LiMoO_2 + Li_{Li} \rightarrow Mo_{Li}^{\cdot\cdot} + O''_i + Li_2O$
Li^+ and Nb^{5+}	Self-Compensation	(iv) $2LiMoO_2 + 2Li_{Li} + Nb_{Nb} \rightarrow Mo_{Li}^{\cdot\cdot} + Mo''_{Nb} + 1.5Li_2O + 0.5Nb_2O_5$
Nb^{5+}	Oxygen Vacancies	(v) $LiMoO_2 + Nb_{Nb} + O_O \rightarrow Mo''_{Nb} + V_O^{\cdot\cdot} + 0.5Li_2O + 0.5Nb_2O_5$
Nb^{5+}	Anti-site (Nb_{Li})	(vi) $2LiMoO_2 + Li_{Li} + 2Nb_{Nb} \rightarrow 2Mo''_{Nb} + Nb_{Li}^{\cdot\cdot\cdot\cdot} + 1.5Li_2O + 0.5Nb_2O_5$
Nb^{5+}	Lithium Vacancies and Anti-site (Nb_{Li})	(vii) $LiMoO_2 + 3Li_{Li} + Nb_{Nb} \rightarrow Mo''_{Nb} + 2V'_{Li} + Nb_{Li}^{\cdot\cdot\cdot\cdot} + 2Li_2O$

According to Figures 2 and 3 for the first and second neighbours with respect to the c axis, the trivalent V^{3+} and Mo^{3+} ions prefer to occupy both the Li and Nb sites according to scheme (iv) which is also observed in other trivalent ions [23–25]. This can be attributed to the similarity between the ionic radius of V^{3+} which is 0.64 Å and Mo^{3+} which is 0.67 Å [40] and that of Li^+ and Nb^{5+}. The ionic radius of Li^+ varies between 0.59 Å and 0.74 Å and Nb^{5+} varies from 0.32 Å to 0.66 Å [40]. All these ionic radii are in relation to the coordination sphere with oxygen atoms.

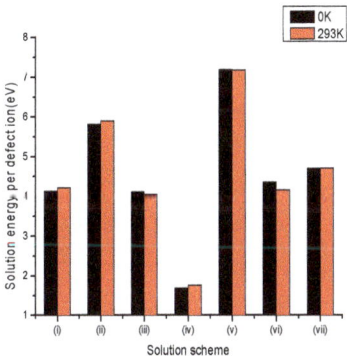

Figure 2. Bar chart of solution energies vs. solution schemes for trivalent dopant (V^{3+}) at the Li and Nb sites, considering the first neighbours in relation to the c axis.

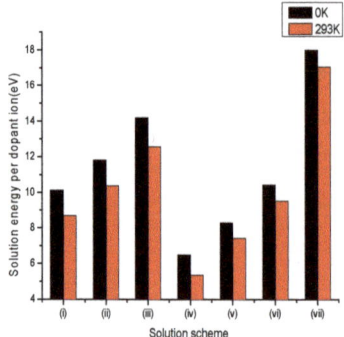

Figure 3. Bar chart of solution energies vs. solution schemes for trivalent dopant (Mo^{3+}) at the Li and Nb sites, considering the first neighbours in relation to the c axis.

3.2.3. Tetravalent Dopants

Like other divalent and trivalent cations, tetravalent V^{4+} and M^{4+} dopant ions can also substitute at either the Li^+ or Nb^{5+} sites. When these ions substitute at the Li^+ and Nb^{5+} site charge compensation is required, and various schemes involving vacancies, interstitials, anti-sites and self-compensation are adopted, as shown in Tables 6 and 7.

Table 6. Types of defects considered due to M = V^{4+} incorporation in $LiNbO_3$.

Site	Charge Compensation	Reaction
Li^+	Lithium Vacancies Niobium Vacancies Oxygen Interstitial	(i) $MO_2 + 4Li_{Li} \rightarrow M_{Li} + 3V'_{Li} + 2Li_2O$ (ii) $5MO_2 + 5Li_{Li} + 3Nb_{Nb} \rightarrow 5M_{Li} + 3V''''_{Nb} + 2.5Li_2O + 1.5Nb_2O_5$ (iii) $2MO_2 + 2Li_{Li} \rightarrow 2M_{Li} + 3O''_i + Li_2O$
Li^+ and Nb^{5+}	Self-Compensation	(iv) $4MO_2 + Li_{Li} + 3Nb_{Nb} \rightarrow M_{Li} + 3M'_{Nb} + 0.5Li_2O + 1.5Nb_2O_5$
Nb^{5+}	Anti-site (Nb_{Li}) Lithium Vacancies and Anti-site (Nb_{Li})	(v) $4MO_2 + Li_{Li} + 4Nb_{Nb} \rightarrow 4M'_{Nb} + Nb_{Li} + 0.5Li_2O + 1.5Nb_2O_5$ (vi) $MO_2 + 4Li_{Li} + Nb_{Nb} \rightarrow M'_{Nb} + 3V'_{Li} + Nb_{Li} + 2Li_2O$
	Oxygen Vacancies	(vii) $2MO_2 + 3Li_{Li} + 2Nb_{Nb} \rightarrow 2M'_{Nb} + 2V'_{Li} + Nb_{Li} + Li_2O + LiNbO_3$ (viii) $3MO_2 + 2Li_{Li} + 3Nb_{Nb} \rightarrow 3M'_{Nb} + V'_{Li} + Nb_{Li} + LiNbO_3$ (ix) $2MO_2 + 2Nb_{Nb} + O_O \rightarrow 2M'_{Nb} + V''_O + Nb_2O_5$

Table 7. Types of defects considered due to M = Mo^{4+} incorporation in $LiNbO_3$.

Site	Charge Compensation	Reaction
Li^+	Lithium Vacancies Niobium Vacancies Oxygen Interstitial	(i) $Li_2MoO_3 + 4Li_{Li} \rightarrow Mo_{Li} + 3V'_{Li} + 3Li_2O$ (ii) $5Li_2MoO_3 + 5Li_{Li} + 3Nb_{Nb} \rightarrow 5Mo_{Li} + 3V''''_{Nb} + 7.5Li_2O + 1.5Nb_2O_5$ (iii) $2Li_2MoO_3 + 2Li_{Li} \rightarrow 2Mo_{Li} + 3O''_i + 3Li_2O$
Li^+ and Nb^{5+}	Self-Compensation	(iv) $4Li_2MoO_3 + Li_{Li} + 3Nb_{Nb} \rightarrow Mo_{Li} + 3Mo'_{Nb} + 4.5Li_2O + 1.5Nb_2O_5$
Nb^{5+}	Anti-site (Nb_{Li}) Lithium Vacancies and Anti-site (Nb_{Li})	(v) $4Li_2MoO_3 + Li_{Li} + 4Nb_{Nb} \rightarrow 4Mo'_{Nb} + Nb_{Li} + 4.5Li_2O + 1.5Nb_2O_5$ (vi) $Li_2MoO_3 + 4Li_{Li} + Nb_{Nb} \rightarrow Mo'_{Nb} + 3V'_{Li} + Nb_{Li} + 3Li_2O$
	Oxygen Vacancies	(vii) $2Li_2MoO_3 + 3Li_{Li} + 2Nb_{Nb} \rightarrow 2Mo'_{Nb} + Nb_{Li} + 2V'_{Li} + 3Li_2O + LiNbO_3$ (viii) $3Li_2MoO_3 + 2Li_{Li} + 3Nb_{Nb} \rightarrow 3Mo'_{Nb} + Nb_{Li} + V'_{Li} + 3Li_2O + 2LiNbO_3$ (ix) $2Li_2MoO_3 + 2Nb_{Nb} + O_O \rightarrow 2Mo'_{Nb} + V''_O + 2Li_2O + Nb_2O_5$

The results obtained from these calculations are given in Figures 4 and 5. By inspecting these figures, it can be seen that the tetravalent cation V^{4+} prefers to be incorporated at the Li^+ and Nb^{5+} sites through scheme (iv), while the Mo^{4+} ion prefers to be incorporated at the niobium site compensated by an oxygen vacancy according to scheme (ix). Similar to the divalent and trivalent dopants, this preference is related to the proximity with the ionic radii of Li^+ and Nb^{5+}.

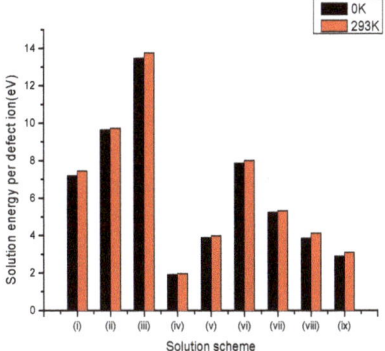

Figure 4. Bar chart of solution energies vs. solution schemes for tetravalent dopant (V^{4+}) at the Li and Nb sites, considering the first neighbours in relation to the c axis.

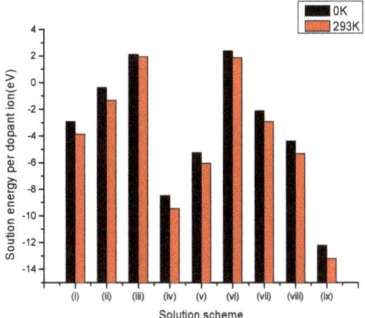

Figure 5. Bar chart of solution energies vs. solution schemes for tetravalent dopant (Mo^{4+}) at the Li and Nb sites, considering the first neighbours in relation to the c axis.

3.2.4. Pentavalent Dopants

For the pentavalent dopants V^{5+} and Mo^{5+}, no charge compensation is required for the substitution at the Nb^{5+} host site, but it is required when the substitution is at the Li^+ host site, as shown in Tables 8 and 9.

Table 8. Types of defects considered due to M = V^{5+} incorporation in $LiNbO_3$.

Site	Charge Compensation	Reaction
Li^+	Lithium Vacancies	(i) $0.5M_2O_5 + 5Li_{Li} \to M_{Li}^{\cdot\cdot\cdot\cdot} + 4V'_{Li} + 2.5Li_2O$
	Niobium Vacancies	(ii) $2.5M_2O_5 + 5Li_{Li} + 5Nb_{Nb} \to 5M_{Li}^{\cdot\cdot\cdot\cdot} + 4V''''_{Nb} + 2.5Li_2O + 2Nb_2O_5$
	Oxygen Interstitial	(iii) $0.5M_2O_5 + Li_{Li} \to M_{Li}^{\cdot\cdot\cdot\cdot} + 2O''_i + 0.5Li_2O$
Nb^{5+}	No Charge Compensation	(iv) $0.5M_2O_5 + Nb_{Nb} \to M_{Nb} + 0.5Nb_2O_5$

Table 9. Types of defects considered due to Mo^{5+} incorporation in $LiNbO_3$.

Site	Charge Compensation	Reaction
Li^+	Lithium Vacancies	(i) $Li_3MoO_4 + 5Li_{Li} \to Mo_{Li}^{\cdot\cdot\cdot\cdot} + 4V'_{Li} + 4Li_2O$
	Niobium Vacancies	(ii) $5Li_3MoO_4 + 5Li_{Li} + 4Nb_{Nb} \to 5Mo_{Li}^{\cdot\cdot\cdot\cdot} + 4V''''_{Nb} + 10Li_2O + 2Nb_2O_5$
	Oxygen Interstitial	(iii) $Li_3MoO_4 + Li_{Li} \to Mo_{Li}^{\cdot\cdot\cdot\cdot} + 2O''_i + 2Li_2O$
Nb^{5+}	No Charge Compensation	(iv) $Li_3MoO_4 + Nb_{Nb} \to Mo_{Nb} + 1.5Li_2O + 0.5Nb_2O_5$

The solution energies for the pentavalent (V^{5+}) and (Mo^{5+}) dopants with different charge compensation mechanisms were evaluated and plotted as a function of the reaction scheme. Based on the lowest energy value, it seems that the incorporation of pentavalent (V^{5+}) and (Mo^{5+}) ions at an Nb site is energetically more favourable than at an Li site, according to scheme (iv) as shown in Figures 6 and 7 at temperatures 0 K and 293 K. This can be attributed to the similarity between the charge of the V^{5+} and Mo^{5+} ions and the Nb^{5+} host, which can contribute to a small deformation in the lattice and consequently a lower solution energy. Experimental results by Kong et al. [17] and Tian et al. [16] show that substitution occurs at the Nb^{5+} site.

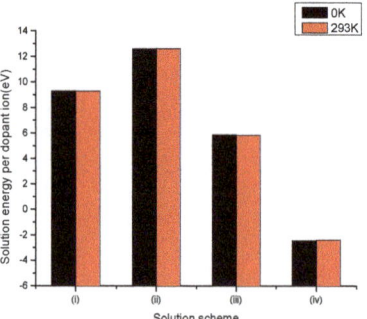

Figure 6. Bar chart of solution energies vs. solution schemes for pentavalent dopant (V^{5+}) at the Li and Nb sites, considering the first neighbours in relation to the c axis.

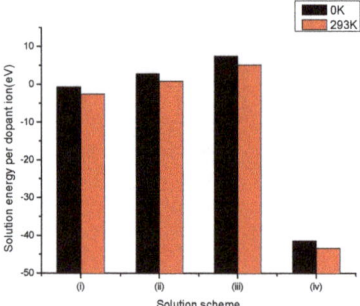

Figure 7. Bar chart of solution energies vs. solution schemes for pentavalent dopant (Mo^{5+}) at the Li and Nb sites, considering the first neighbours in relation to the c axis.

3.2.5. Hexavalent Dopants

For the hexavalent dopant Mo^{6+}, as with the pentavalent ions, there is no self-compensation mechanism and charge compensation schemes are possible when replacing Li and Nb in the $LiNbO_3$ matrix as shown in Table 10.

Table 10. Types of defects considered due to Mo^{6+} incorporation in $LiNbO_3$.

Site	Charge Compensation	Reaction
Li^+	Lithium Vacancies	(i) $Li_2MoO_4 + 6Li_{Li} \rightarrow Mo_{Li}^{\cdot} + 5V'_{Li} + 4Li_2O$
	Niobium Vacancies	(ii) $Li_2MoO_4 + Li_{Li} + Nb_{Nb} \rightarrow Mo_{Li}^{\cdot} + V'''''_{Nb} + 1.5Li_2O + 0.5Nb_2O_5$
	Oxygen Interstitial	(iii) $2Li_2MoO_4 + 2Li_{Li} \rightarrow 2Mo_{Li}^{\cdot} + 5O''_i + 3Li_2O$
Nb^{5+}	Lithium Vacancies	(iv) $Li_2MoO_4 + Li_{Li} + Nb_{Nb} \rightarrow Mo_{Nb} + V'_{Li} + 1.5Li_2O + 0.5Nb_2O_5$
	Niobium Vacancies	(v) $5Li_2MoO_4 + 6Nb_{Nb} \rightarrow 5Mo_{Nb} + V'''''_{Nb} + 5Li_2O + 3Nb_2O_5$
	Oxygen Interstitial	(vi) $2Li_2MoO_4 + 2Nb_{Nb} \rightarrow 2Mo_{Nb} + O''_i + 2Li_2O + Nb_2O_5$

The solution energies for the hexavalent (Mo^{6+}) dopants with different charge-compensation mechanisms were evaluated and plotted as a function of the reaction scheme. Based on the lowest energy value, it seems that the incorporation of hexavalent (Mo^{6+}) ions at an Nb site is energetically more favourable than at an Li site, according to scheme (iv) as shown in Figure 8 at temperatures 0 K and 293 K. This can be attributed to the similarity between the ionic radii of Mo^{6+} ions and the Nb^{5+} host site (0.32–0.71 Å) [40]. The ionic radii of Mo^{6+}, taking into account the coordination number, vary between 0.42 and 0.67 Å [40], and the small difference between the Mo^{6+} dopant ions and Nb^{5+} ions can contribute to a small deformation in the lattice and consequently a lower solution energy. This result reveals that global trends of dopant solution energies are controlled by the combination of dopant ion size [40] and its electrostatic interactions, demonstrating that there is a relation between the energetically preferred site and the types of defect mechanisms involved in the doping process. Experimental results from Kong et al. [17] and Zhu et al. [41] show that substitution occurs at the Nb^{5+} site.

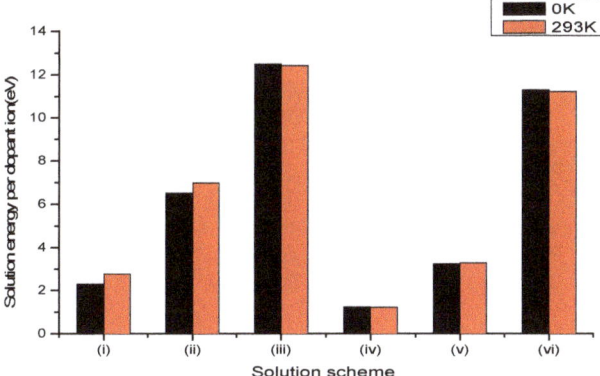

Figure 8. Bar chart of solution energies vs. solution schemes for hexavalent dopant (Mo^{6+}) at the Li and Nb sites, considering the first neighbours in relation to the c axis.

In all cases, the energy involved in doping was obtained by calculating the solution energy, which includes all terms of the thermodynamic cycle involved in the solution process. For example, the solution energy, E_{sol}, corresponding to the incorporation of V^{2+} at the Li$^+$ site (second equation in Table 3) is given by:

$$E_{Sol} = E_{Def}(5M_{Li} + V'''''_{Nb}) + 2.5E_{Latt}(Li_2O) + 0.5E_{Latt}(Nb_2O_5) - 5E_{Latt}(MO) \qquad (2)$$

where the E_{latt} and E_{Def} terms are lattice energies and defect energy.

All energies were normalised by the number of dopants, i.e., the solution energy is divided by the number of dopants involved. For example, for scheme (ii) of Table 3, the energy must be divided by five, since five lithium sites are occupied. This is done because the number of dopants varies for each mechanism. Lattice energies, E_{latt}, required to calculate the solution energies are given in Table 11.

Table 11. Lattice energies used in the solution energy calculations (eV).

Compound	Lattice Energy 0 K	Lattice Energy 293 K
$LiNbO_3$	−174.45	−174.66
Li_2O	−33.16	−32.92
Nb_2O_5	−314.37	−313.39
VO	−22.06	−22.07
V_2O_3	−124.37	−124.39
VO_2	−111.54	−111.57
V_2O_5	−315.65	−274.18
$LiMoO_2$	−98.07	−97.09
Li_2MoO_3	−150.38	−149.10
Li_3MoO_4	−181.28	−178.88
Li_2MoO_4	−234.06	−234.12

3.2.6. Summary of Results for Vanadium and Molybdenum Dopants in $LiNbO_3$

In this sub-section, the results presented in the last five subsections are summarised.

Divalent dopants: the calculations predict that, for V^{2+}, self-compensation (simultaneous doping at lithium and niobium sites) and doping at the lithium site with lithium vacancy compensation are most likely. It is noted that $V^{2+}{}_{Li}$ defects have been observed experimentally [18].

Trivalent dopants: both V^{3+} and Mo^{3+} ions are predicted to self-compensate. Experimental data from [18] support V^{3+} doping at the lithium site, as with V^{2+}.

Tetravalent dopants: here, different behaviour is predicted for vanadium and molybdenum. V^{4+} is predicted to self-compensate, while Mo^{4+} is predicted to occupy a niobium site with oxygen vacancy charge compensation. Again, [18] suggests that V^{4+} can dope at a lithium site.

Pentavalent dopants: both V^{5+} and Mo^{5+} are predicted to dope at the niobium site (no charge compensation is needed), agreeing with experimental results [16,17].

Hexavalent dopants: Mo^{6+} is predicted to dope at the niobium site, with charge compensation by lithium vacancy formation. The occupation of the niobium site is supported by experimental data [16,17,19].

4. Conclusions

This paper has presented a computational study of VO, V_2O_3, VO_2 and V_2O_5 as well as $LiMoO_2$, Li_2MoO_3, Li_3MoO_4 and Li_2MoO_4 structures doped into $LiNbO_3$. New interatomic potential parameters for VO, V_2O_3, VO_2 and V_2O_5 as well as $LiMoO_2$, Li_2MoO_3, Li_3MoO_4 and Li_2MoO_4 have been developed. It was found that divalent (V^{2+}), trivalent (V^{3+}, Mo^{3+}) and tetravalent (V^{4+}) ions are more favourably incorporated at the Li and Nb sites through the self-compensation mechanism. The tetravalent (Mo^{4+}) ion is more favourably incorporated at the niobium site, compensated by an oxygen vacancy. The pentavalent ions (V^{5+}, Mo^{5+}) and hexavalent (Mo^{6+}) ion are more favourably incorporated at the Nb site, and the lowest energy schemes involve, respectively, no charge compensation, and for the Mo^{6+} ion, charge compensation with lithium vacancy. This is shown to be consistent with some experimental data, although future calculations involving finite V^{5+} and Mo^{6+} concentrations will be carried out to investigate this further.

Finally, to summarise, in this paper we have looked in detail at vanadium and molybdenum dopants in various charge states in $LiNbO_3$, and through the use of solution energies, identified the energetically favoured sites and charge compensation mechanisms, while comparing the results with available experimental and theoretical work in this field.

Author Contributions: Conceptualisation, R.M.A.; Data curation, R.M.A. and E.F.d.S.M.; Formal analysis, R.A.J.; Supervision, M.E.G.V. and R.A.J.; Validation, M.E.G.V. and R.A.J.; Writing—original draft, R.M.A.; Writing—review & editing, M.E.G.V. and R.A.J. All authors have read and agreed to the published version of the manuscript.

Funding: This research received no external funding.

Acknowledgments: The authors would like to thank the peer reviewers, whose detailed comments have undoubtedly led to major improvements to this paper.

Conflicts of Interest: The authors declare no conflict of interest.

References

1. Mandula, G.; Rupp, R.A.; Balaskó, M.; Kovács, L. Decay of photorefractive gratings in LiNbO$_3$:Fe by neutron irradiation. *Appl. Phys. Lett.* **2005**, *86*, 141107. [CrossRef]
2. Ionita, I.; Jaque, F. Photoconductivity and electron mobility in LiNbO$_3$ co-doped with Cr^{3+} and MgO. *Opt. Mater.* **1998**, *10*, 171–173. [CrossRef]
3. Kaczmarec, S.M.; Bodziony, T. Low symmetry centers in LiNbO$_3$ with Yb and Er. *J. Non-Cryst. Solids* **2008**, *354*, 4202–4210. [CrossRef]
4. Kokanyan, E.P.; Razzari, L.; Cristiani, I.; Degiorgio, V.; Gruber, J.B. Reduced photorefraction in hafnium-doped single-domain and periodically poled lithium niobate crystals. *Appl. Phys. Lett.* **2004**, *84*, 1880. [CrossRef]
5. Corradi, G.; Meyer, M.; Kovács, L.; Polgár, K. Gap levels of Ti^{3+} on Nb or Li sites in LiNbO$_3$:(Mg):Ti crystals and their effect on charge transfer processes. *Appl. Phys. B* **2004**, *78*, 607–614. [CrossRef]
6. Cantelar, E.; Quintanilla, M.; Pernas, P.L.; Torchia, G.A.; Lifante, G.; Cussó, F. Polarized emission and absorption cross-section calculation in LiNbO$_3$:Tm^{3+}. *J. Lumin.* **2008**, *128*, 988–991. [CrossRef]
7. Shura, J.W.; Shina, T.I.; Leea, S.M.; Baekb, S.W.; Yoon, D.H. Photoluminescence properties of Nd: LiNbO$_3$ co-doped with ZnO fiber single crystals grown by micro-pulling-down method. *Mater. Sci. Eng. B* **2003**, *105*, 16–19. [CrossRef]
8. Li, S.; Liu, S.; Kong, Y.; Deng, D.; Gao, G.; Li, Y.; Gao, H.; Zhang, L.; Hang, Z.; Chen, S.; et al. The optical damage resistance and absorption spectra of LiNbO$_3$:Hf crystals. *J. Phys. Condens. Matter* **2006**, *18*, 3527–3534. [CrossRef]
9. Hesselink, L.; Orlov, S.S.; Liu, A.; Akella, A.; Lande, D.; Neurgaonkar, R.R. Photorefractive materials for nonvolatile volume holographic data storage. *Science* **1998**, *282*, 1089–1094. [CrossRef]
10. Camarillo, E.; Murrieta, H.; Hernandez, J.M.; Zoilo, R.; Flores, M.C.; Han, T.P.J.; Jaque, F. Optical properties of LiNbO$_3$:Cr crystals co-doped with germanium oxide. *J. Lumin.* **2008**, *128*, 747–750. [CrossRef]
11. Luo, S.; Meng, Q.; Wang, J.; Sun, X. Effect of In^{3+} concentration on the photorefraction and scattering properties in In: Fe:Cu:LiNbO$_3$ crystals at 532 nm wavelength. *Opt. Commun.* **2016**, *358*, 198–201. [CrossRef]
12. Nie, Y.; Wang, R.; Wang, B. Growth and holographic storage properties of In:Ce:Cu:LiNbO$_3$ crystal. *Mater. Chem. Phys.* **2007**, *102*, 281–283. [CrossRef]
13. Zhen, X.H.; Li, H.T.; Sun, Z.J.; Ye, S.J.; Zhao, L.C.; Xu, Y.H. Holographic properties of double-doped Zn:Fe:LiNbO$_3$ crystals. *Mater. Lett.* **2004**, *58*, 1000–1002. [CrossRef]
14. Wei, Z.; Naidong, Z.; Qingquan, L. Growth and Holographic Storage Properties of Sc, Fe Co-Doped Lithium Niobate Crystals. *J. Rare Earth* **2007**, *25*, 775–778. [CrossRef]
15. Xu, C.; Leng, X.; Xu, L.; Wen, A.; Xu, Y. Enhanced nonvolatile holographic properties in Zn, Ru and Fe co-doped LiNbO$_3$ crystals. *Opt. Commun.* **2012**, *285*, 3868–3871. [CrossRef]
16. Tian, T.; Kong, Y.; Liu, S.; Li, W.; Wu, L.; Chen, S.; Xu, J. The photorefraction of molybdenum-doped lithium niobate crystals. *Opt. Lett.* **2012**, *37*, 2679–2681. [CrossRef]
17. Kong, Y.; Liu, S.; Xu, J. Recent Advances in the Photorefraction of Doped Lithium Niobate Crystals. *Materials* **2012**, *5*, 1954–1971. [CrossRef]
18. Saeed, S.; Zheng, D.; Liu, H.; Xue, L.; Wang, W.; Zhu, L.; Hu, M.; Liu, S.; Chen, S.; Zhang, L.; et al. Rapid response of photorefraction in vanadium and magnesium co-doped lithium niobate. *J. Phys. D Appl. Phys.* **2019**, *52*, 405303. [CrossRef]
19. Xue, L.; Liu, H.; Zheng, D.; Saeed, S.; Wang, X.; Tian, T.; Zhu, L.; Kong, Y.; Liu, S.; Chen, S.; et al. The Photorefractive Response of Zn and Mo Codoped LiNbO$_3$ in the Visible Region. *Crystals* **2019**, *9*, 228. [CrossRef]

20. Fan, Y.; Li, L.; Li, Y.; Sun, X.; Zhao, X. Hybrid density functional theory study of vanadium doping in stoichiometric and congruent LiNbO$_3$. *Phys. Rev. B* **2019**, *99*, 035147. [CrossRef]
21. Wang, W.; Liu, H.; Zheng, D.; Kong, Y.; Zhang, L.; Xu, J. Interaction between Mo and intrinsic or extrinsic defects of Mo doped LiNbO$_3$ from first-principles calculations. *J. Phys. Condens. Matter* **2020**, *32*, 255701. [CrossRef] [PubMed]
22. Jackson, R.A.; Valerio, M.E.G. A new interatomic potential for the ferroelectric and paraelectric phases of LiNbO$_3$. *J. Phys. Condens. Matter* **2005**, *17*, 837. [CrossRef]
23. Araujo, R.M.; Lengyel, K.; Jackson, R.A.; Valerio, M.E.G.; Kovacs, L. Computer modelling of intrinsic and substitutional defects in LiNbO$_3$. *Phys. Status Solidi* **2007**, *4*, 1201–1204.22. [CrossRef]
24. Araujo, R.M.; Lengyel, K.; Jackson, R.A.; Kovacs, L.; Valerio, M.E.G. A computational study of intrinsic and extrinsic defects in LiNbO$_3$. *J. Phys. Condens. Matter* **2007**, *19*, 046211. [CrossRef]
25. Araujo, R.M.; Valerio, M.E.G.; Jackson, R.A. Computer modelling of trivalent metal dopants in lithium niobite. *J. Phys. Condens. Matter* **2008**, *20*, 035201. [CrossRef]
26. Araujo, R.M.; Valerio, M.E.G.; Jackson, R.A. Computer simulation of metal co-doping in lithium niobate. *Proc. R. Soc. A* **2014**, *470*, 0406. [CrossRef]
27. Araujo, R.M.; Valerio, M.E.G.; Jackson, R.A. Computer Modelling of Hafnium Doping in Lithium Niobate. *Crystals* **2018**, *8*, 123. [CrossRef]
28. Mott, N.F.; Littleton, M.J. Conduction in polar crystals. Electrolytic conduction in solid salts. *Trans. Faraday Soc.* **1938**, *34*, 485–499. [CrossRef]
29. Sanders, M.J.; Leslie, M.; Catlow, C.R.A. Interatomic potentials for SiO$_2$. *J. Chem. Soc. Chem. Commun.* **1984**, 1271–1273. [CrossRef]
30. Dick, B.J.; Overhauser, A.W. Theory of the dielectric constants of alkali halide crystals. *Phys. Rev.* **1958**, *112*, 90. [CrossRef]
31. Taylor, D. Thermal expansion data. I: Binary oxides with the sodium chloride and wurtzite structures, MO. *Trans. J. Br. Ceram. Soc.* **1984**, *83*, 5–9.
32. Luedtke, T.; Weber, D.; Schmidt, A.; Mueller, A.; Reimann, C.; Becker, N.; Bredow, T.; Dronskowski, R.; Ressler, T.; Lerch, M. Synthesis and characterization of metastable transition metal oxides and oxide nitrides. *Z. Fuer Krist. Cryst. Mater.* **2017**, *232*, 3–14. [CrossRef]
33. McWhan, D.B.; Marezio, M.; Remeika, J.P.; Dernier, P.D. X-ray diffraction study of metallic VO$_2$. *Phys. Rev. B Solid State* **1974**, *10*, 490–495. [CrossRef]
34. Balog, P.; Orosel, D.; Cancarevic, Z.; Schoen, C.; Jansen, M. V$_2$O$_5$ phase diagram revisited at high pressures and high temperatures. *J. Alloy. Compd.* **2007**, *429*, 87–98. [CrossRef]
35. Aleandri, L.E.; McCarley, R.E. Hexagonal lithium molybdate, LiMoO$_2$: A close-packed layered structure with infinite molybdenum-molybdenum-bonded sheets. *Inorg. Chem.* **1988**, *27*, 1041–1044. [CrossRef]
36. Hibble, S.J.; Fawcett, I.D.; Hannon, A.C. Structure of Two Disordered Molybdates, Li$_2$Mo^{IV}O$_3$ and Li$_4$Mo$_3^{IV}$O$_8$, from Total Neutron Scattering. *Acta Crystallogr. Sect. B Struct. Sci.* **1997**, *53*, 604–612. [CrossRef]
37. Mikhailova, D.; Voss, A.; Oswald, S.; Tsirlin, A.A.; Schmidt, M.; Senyshyn, A.; Eckert, J.; Ehrenberg, H. Lithium Insertion into Li$_2$MoO$_4$: Reversible Formation of (Li$_3$Mo)O$_4$ with a Disordered Rock-Salt Structure. *Chem. Mater.* **2015**, *27*, 4485–4492. [CrossRef]
38. Kolitsch, U. The crystal structures of phenacite-type Li$_2$(MoO$_4$), and scheelite-type LiY(MoO$_4$)$_2$ and LiNd(MoO$_4$)$_2$. *Z. Fuer Krist.* **2001**, *216*, 449–454. [CrossRef]
39. Kröger, F.A.; Vink, H.J. The origin of the fluorescence in self-activated ZnS, CdS, and ZnO. *J. Chem. Phys.* **1954**, *22*, 250. [CrossRef]
40. Shannon, R.D.; Prewitt, C.T. Revised values of effective ionic radii. *Acta Crystallogr. Sect. B Struct. Crystallogr. Cryst. Chem.* **1969**, *26*, 1046–1048. [CrossRef]
41. Zhu, L.; Zheng, D.; Saeed, S.; Wang, S.; Liu, H.; Kong, Y.; Liu, S.; Chen, S.; Zhang, L.; Xu, J. Photorefractive Properties of Molybdenum and Hafnium Co-Doped LiNbO$_3$. *Crystals* **2018**, *8*, 322. [CrossRef]

© 2020 by the authors. Licensee MDPI, Basel, Switzerland. This article is an open access article distributed under the terms and conditions of the Creative Commons Attribution (CC BY) license (http://creativecommons.org/licenses/by/4.0/).

Article

Improvement on Thermal Stability of Nano-Domains in Lithium Niobate Thin Films

Yuejian Jiao, Zhen Shao, Sanbing Li, Xiaojie Wang, Fang Bo, Jingjun Xu and Guoquan Zhang *

The MOE Key Laboratory of Weak-Light Nonlinear Photonics, School of Physics and TEDA Applied Physics Institute, Nankai University, Tianjin 300457, China; jiaoyuejian@mail.nankai.edu.cn (Y.J.); 2120160179@mail.nankai.edu.cn (Z.S.); 2120180199@mail.nankai.edu.cn (S.L.); xjw@nankai.edu.cn (X.W.); bofang@nankai.edu.cn (F.B.); jjxu@nankai.edu.cn (J.X.)
* Correspondence: zhanggq@nankai.edu.cn

Received: 23 December 2019; Accepted: 22 January 2020; Published: 30 January 2020

Abstract: We present a simple and effective way to improve the thermal stability of nano-domains written with an atomic force microscope (AFM)-tip voltage in a lithium niobate film on insulator (LNOI). We show that nano-domains in LNOI (whether in the form of stripe domains or dot domains) degraded, or even disappeared, after a post-poling thermal annealing treatment at a temperature on the order of ~100 °C. We experimentally confirmed that the thermal stability of nano-domains in LNOI is greatly improved if a pre-heat treatment is carried out for LNOI before the nano-domains are written. This thermal stability improvement of nano-domains is mainly attributed to the generation of a compensating space charge field parallel to the spontaneous polarization of written nano-domains during the pre-heat treatment process.

Keywords: thermal stability; nano-domain; LNOI; pre-heat treatment

1. Introduction

Lithium niobate ($LiNbO_3$), one of the most versatile ferroelectric materials, has been widely studied due to its excellent performance on electro-optic modulation [1,2], acousto-optic modulation [3] and nonlinear optics [4,5]. Recently, the technique of lithium niobate film on insulators (LNOI) [6–8] has attracted much attention for its potential applications in integrated devices. Numerous novel optical elements based on LNOI have been reported, including photonic crystals [9], high-Q microresonators [10], ridge waveguides [11], and hybrid lightwave circuits [12].

By applying a polarization reversal voltage, via an atomic force microscope (AFM)-tip, domain reversal and domain patterning can be realized in lithium niobate thin films. Based on this technique, Gainutdinov et al. [13] reported that the size and shape of domain patterns can be precisely controlled, enabling the realization of periodically poled lithium niobate (PPLN) with period of hundreds of nanometers. PPLN films can be used for quasi-phase-matching (QPM) devices, such as PPLN microcavities [14] and PPLN waveguides [15], to achieve frequency conversion. Obviously, the stability of written domains is very important for LNOI-based applications such as PPLN microcavities, PPLN waveguides, and nonvolatile ferroelectric domain memories [16–19].

Several groups have studied the thermal stability of domains in various ferroelectric materials, such as Rb-doped $KTiOPO_4$ [20], $LiTaO_3$ [21], $Pb(Zr_{0.4}Ti_{0.6})O_3$ [22], and $LiNbO_3$ [23,24]. They reported that the ferroelectric domains would degrade, or even disappear, after heat treatment. Moreover, Shao et al. [25] reported that the domain structures fabricated on LNOI are unstable even at room temperature. Obviously, such instability would prevent the ferroelectric domains from applications where the device temperature will rise due to light absorption or due to high temperature environments.

In this paper, we propose a simple and effective method to improve the thermal stability of nano-domains in lithium niobate thin films. We confirmed that nano-domains written in LNOI by applying an AFM-tip voltage were unstable at high temperatures in the order of ~100 °C. However, we found that the domain stability can be significantly improved if the LNOI sample experiences a pre-heat treatment before the nano-domain fabrication process. The underlying mechanism was also discussed.

2. Materials and Methods

The schematic experimental setup is shown in Figure 1, in which the structure of the LNOI sample used in our experiments is also clearly shown. The LNOI sample was composed of a 300-nm thick +Z-cut ion-sliced $LiNbO_3$ thin film, a 100-nm thick Cr thin film, a 2-μm thick SiO_2 layer, and a 500-μm thick $LiNbO_3$ substrate, which were all layered or bonded to one another in sequence. The 100-nm thick Cr layer served as a bottom electrode when an AFM-tip voltage was applied on the top 300-nm thick $LiNbO_3$ thin film. Here, different metals may be used as the bottom electrode and different metal-lithium-niobate interfaces may have an effect on the domain poling process, but this is not the main topic of the current paper and will not be explored here.

In the experiments, the top $LiNbO_3$ film was poled directly by applying a DC voltage through an AFM conductive probe tip, contacting the film top surface with the Cr layer being grounded. The dot domains were written under the AFM-tip voltage step by step, and the stripe domain patterns were written using a raster lithography method with graphic templates. The reversed domain structures were characterized by using piezoresponse force microscope (PFM), a versatile and powerful method to image domain structures with nano-size features. The tip radius, R, and the resonance frequency, f_R, of the pt-coated Si probe tip used in the experiments were $R = 20$ nm and $f_R = 100$ kHz, respectively. All AFM and PFM experiments were carried out with an MFP-3D Infinity atomic force microscope (Asylum Research, Goleta, CA, USA).

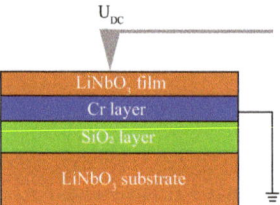

Figure 1. Schematic diagram of nano-domain writing in lithium niobate film on insulators (LNOI) under an atomic force microscope (AFM)-tip voltage.

The thermal heat treatments, including the post-poling annealing treatment after the domain writing process and the pre-heat treatment with the virgin LNOI sample without domain structure, were carried out by using an electric drying oven. The sample was heated to a temperature ranging from 90 °C to 210 °C in air, with a heating rate of 5 °C/min from room temperature, and then maintained at the high temperature for a certain time. After that, the sample was moved out from the drying oven and cooled down naturally to room temperature in air with a cooling rate of ~20 °C/min. Note that no oxidation or reduction effect was observed in lithium niobate thin films during the thermal annealing treatment at a temperature of the order of 100 °C.

3. Results

3.1. Thermal Stability of Nano-Domains in Lithium Niobate Thin Films

To begin with, we will explore the thermal stability of nano-domains in lithium niobate thin films without any pre-heat treatment in this part. As QPM devices and ferroelectric domain memory are two important potential applications for domain structures, both stripe domains and dot domains were fabricated and studied. Here, the stripe domains were fabricated using a raster lithography method with an AFM-tip voltage of 35 V. The rate of lithography was fixed at $f = 2$ Hz. Periodical stripe domains with a fixed period of 1 µm and an averaged stripe length of ~4 µm but with different stripe widths of w = 396 nm, 205 nm, and 156 nm, were fabricated. The PFM images of these as-written stripe domains were measured, and the results are shown in Figure 2a–c, respectively.

We confirmed experimentally that these as-written stripe domians were stable at room temperature, and no degradation was observed even for several days. Then, the stripe domains were thermally annealed at a high temperature T = 120 °C for t = 1 h and then cooled down naturally to room temperature in air again. For comparison, the PFM images of the stripe domains after the thermal annealing treatment are shown in Figure 2d–f, respectively. The stripe domains were significantly degraded in both width and length dimensions after the thermal annealing treatment.

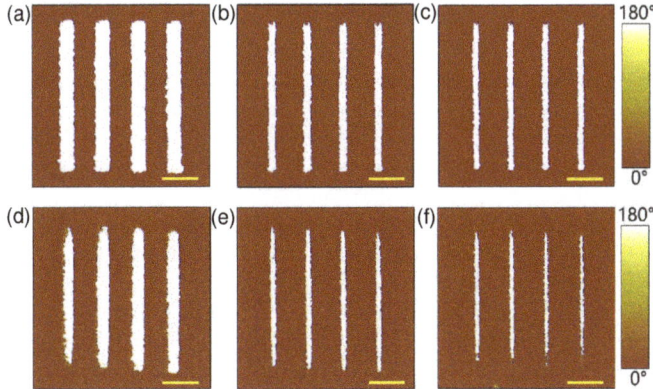

Figure 2. (a–c) Piezoresponse force microscope (PFM) images of as-written stripe domains with different stripe widths before the thermal annealing treatment. (d–f) The corresponding PFM images of stripe domains after the thermal annealing treatment at T = 120 °C for 1 h. The length of the stripe domains was set to be ~ 4 µm. The scale bar in all figures is 1 µm.

In addition, dot domains with different diameters were also fabricated by applying different AFM-tip voltages for a fixed time $t_w = 1$ s. Each dot domain was separated from one another by 1 µm in both the horizontal and vertical directions. Figure 3a shows the PFM images of the fabricated dot domains, in which four dot domains in each row were fabricated with the same tip voltage. These voltages were, from the bottom up, 40 V, 45 V, 50 V, and 55 V. The averaged diameter of the as-written dot domains in each row was measured to be 215 nm, 255 nm, 294 nm, and 333 nm, respectively. Here, the diameter D of a dot domain was estimated by equaling the area of the dot domain to a circle with a diameter D.

These as-written dot domains were also stable at room temperature. After that, the dot domains were annealed thermally at a high temperature T = 120 °C for one hour, and then the dot domains were cooled down to room temperature in air. Again, the PFM images of the dot domains after thermal annealing treatment were measured, and the results are shown in Figure 3b. It is evident that the dot domains are significantly degraded and even disappear for those small dot domains. This observed thermal instability is likely detrimental for practical applications such as QPM devices and ferroelectric domain memory devices.

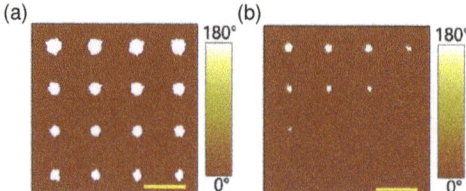

Figure 3. (**a**) PFM images of as-written dot domains with different diameters before the thermal annealing treatment. (**b**) The corresponding PFM images of dot domains after the thermal annealing treatment, at a temperature T = 120 °C for 1 h. Here, the dot domains in each row were fabricated at the same AFM-tip voltage. These voltages were, from the bottom up, 40 V, 45 V, 50 V, and 55 V. The scale bar was 1 µm in both cases.

3.2. Improvement on the Thermal Stability of Nano-Domains in Pre-Heated Lithium Niobate Thin Films

Here, we introduce a simple but effective way to improve the thermal stability of nano-domains in lithium niobate thin films. First, a virgin single-domain sample without any domain structures was put into the electric drying oven to undergo a pre-heat treatment at T_p = 150 °C for 2 h. Then, nano-domains were written with the same tip voltage as those in Figure 2 for stripe domains and in Figure 3 for dot domains. In the experiments, the period of the stripe domains was set to be 1 µm, and the width of the stripe domains was set to be 333 nm, 215 nm, and 137 nm, respectively. The PFM images of these as-written stripe domains were measured and are shown in Figure 4a–c, respectively.

After that, the sample with the stripe domains was thermally annealed at T = 120 °C for 1 h and then cooled down naturally to room temperature in air. The PFM images of the stripe domains were measured again for comparison, after the thermal annealing treatmen, and the results are shown in Figure 4d–f, respectively. As shown in Figure 4, although the stripe domains with a pre-heat treatment also degrade after the post-poling thermal annealing treatment, the degradation is significantly suppressed as compared to the case without the pre-heat treatment.

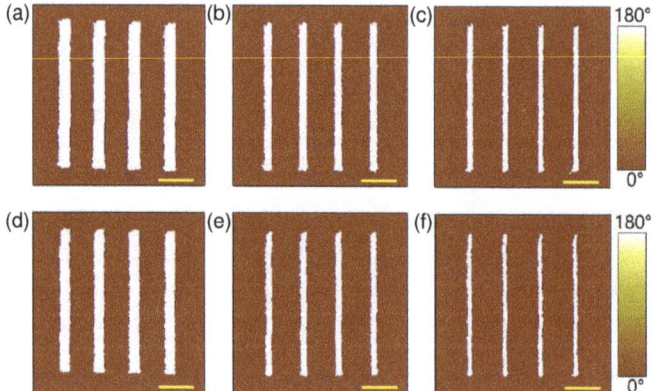

Figure 4. The thermal stability of stripe domains in a sample that underwent a pre-heat treatment at T_p = 150 °C for 2 h. (**a**–**c**) PFM images of as-written stripe domains before the thermal annealing treatment. (**d**–**f**) PFM images of stripe domains after the thermal annealing treatment at 120 °C for 1 h. The length of the stripe domains was set to be ∼ 4 µm. Here, the tip voltage used to fabricate the stripe domains was the same as that in Figure 2. The scale bar is 1 µm in all figures.

The thermal stability of the dot domains in the pre-heat treated samples was also studied. In the experiments, the dot domains were written in the pre-heat treated sample under the same tip voltage and writing time t_w as those in Figure 3. Similarly, the separation distance between the nearest neighboring dot domains was set to be 1 µm in both the horizontal and vertical dimensions, and dot domains with different averaged diameters of 215 nm, 255 nm, 294 nm, and 333 nm were prepared. Again, the diameters of the dot domains were averaged over four dot domains fabricated under the same tip voltage and writing time t_w. Then, the sample with the dot domains underwent the same thermal annealing process as that in Figure 3.

The PFM images of the dot domains before and after the post-poling thermal annealing treatment were measured for comparison, and the results are shown in Figure 5. Compared to the case without pre-heat treatment in Figure 3, the thermal stability of the dot domains in the pre-heat treated samples is significantly improved.

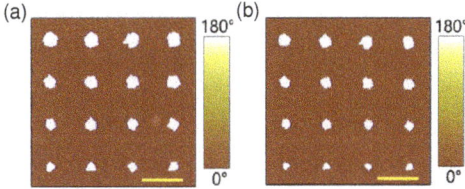

Figure 5. Thermal stability of the dot domains in a sample that underwent a pre-heat treatment at $T_p = 150$ °C for 2 h. (**a**) PFM images of as-written dot domains with different diameters before the thermal annealing treatment. (**b**) PFM images of dot domains after the thermal annealing treatment at 120 °C for 1 h. Here, the dot domains in each row were fabricated at the same AFM-tip voltage as those in Fig. 3. The scale bar is 1 µm in all figures.

4. Discussions

To show quantitatively the improvement on thermal stability of nano-domains in the pre-heat treated samples, we introduced a thermal stability parameter P, defined as $P = S_{remain}/S_{initial}$, where $S_{initial}$ and S_{remain} are the areas of the nano-domains before and after the post-poling thermal annealing treatment. The domain is more stable for a larger P. Table 1 lists the values of the thermal stability parameter P for both stripe domains and dot domains, as shown in Figures 2–5.

In general, as compared to the case without pre-heat treatment, the thermal stability parameter P is much larger for nano-domains in the pre-heat treated samples, indicating that the thermal stability of nano-domains in samples with pre-heat treatment is significantly improved. Note that the length of stripe domains also shrinks, and the length shrinkages were measured to be 0.294 µm, 0.235 µm, and 0.588 µm, in the case without pre-heat treatment, while in the case with pre-heat treatment, the length shrinkages were reduced to be 0.125 µm, 0.121 µm, and 0.093 µm, for stripe domains with widths of 333 nm, 215 nm, and 137 nm, respectively.

Table 1. The thermal stability parameter, P, of nano-domains in samples with or without pre-heat treatment. The condition of pre-heat treatment was $T_p = 150$ °C for 2 hours. The length of the stripe domains was set to be ~4 µm.

Domain Size	Width of Stripe Domains			Diameter of Dot Domains		
	333 nm	215 nm	137 nm	333 nm	255 nm	215 nm
without pre-heat treatment	0.69	0.64	0.52	0.24	0	0
with pre-heat treatment	0.86	0.83	0.79	0.72	0.66	0.56

The dependence of the thermal stability parameter, P, on the post-poling annealing temperature, T, was studied for both stripe and dot domains without pre-heat treatment, and the results are shown in Figure 6. Here, stripe domains with different widths of 372 nm, 196 nm, and 155 nm and dot domains with different diameters of 333 nm, 255 nm, and 215 nm, were prepared. The length of all stripe domains was set to be ~4 μm. In all cases, the post-poling thermal annealing time, t, was set to be one hour. For both stripe domains and dot domains, the thermal stability parameter, P, decreases with the increase of the post-poling annealing temperature, T, and the domain degradation at 120 °C is typical of the representative results within the studied temperature range, which is practically reachable in nano-size photonic structures, such as PPLN microcavities and PPLN ridge waveguides.

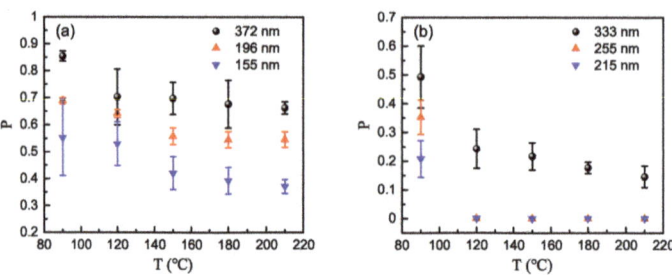

Figure 6. The dependence of the thermal stability parameter, P, on the post-poling annealing temperature T for both stripe domains (**a**) and dot domains (**b**) of various sizes without pre-heat treatment. Here, the post-poling thermal annealing time t was set to be 1 h in all cases, and the length of the stripe domains was set to be ~4 μm.

It has been reported that the domain structures in bulk lithium niobate crystals are stable at temperatures on the order of 100 °C but decay also at a much higher temperature above 600 °C [26,27], indicating that the domain structure in bulk crystal sheets is much more thermally stable when compared to that in lithium niobate thin films.

Furthermore, we studied the dependence of the thermal stability of nano-domains on the experimental pre-heat treatment conditions. In the experiments, pre-heat treatment on virgin single-domain samples was carried out at different high temperatures, T_p, for different time periods, t_p, and then stripe or dot domains with different sizes were fabricated by applying appropriate tip voltages. After that, the nano-domains were thermally annealed at T = 120 °C for 1 h. The PFM images of all nano-domains were measured and the thermal stability parameter P was characterized for each nano-domain.

Figure 7a,b shows the dependence of the thermal stability parameter, P, on the pre-heat temperature, T_p, with t_p = 2 h for the stripe domains and dot domains, with various sizes. P increases with the increase of the pre-heat temperature, T_p, in both the stripe domain and the dot domain cases, indicating that the nano-domains are more thermally stable with higher T_p. Figure 7c,d depicts the dependence of the thermal stability parameter, P, on the pre-heat time, t_p, for the stripe and dot domains of various sizes. Here, the pre-heat treatment temperature was set to be 150 °C for both cases. The thermal stability parameter P is larger with longer pre-heat treatment time, t_p. In addition, the nano-domains with larger sizes are more stable for both cases, as shown in Figure 7.

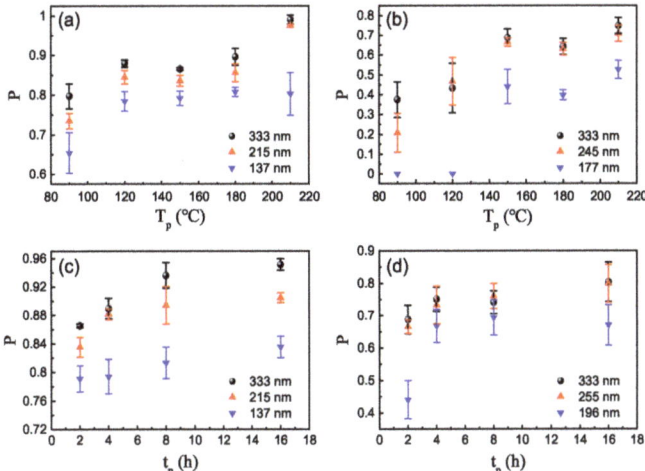

Figure 7. Dependence of the thermal stability parameter, P, on the pre-heat temperature, T_p, (**a,b**) and the pre-heat time, t_p, (**c,d**), for nano-domains with various sizes. Here, (**a,c**) are the results for the stripe domains, while (**b,d**) are the results for the dot domains. The pre-heat time, t_p, was 2 hours for (**a,b**), while the pre-heat temperature, T_p, was set to be 150 °C for (**c,d**). All nano-domains were thermally annealed at T = 120 °C for 1 h.

From the above results, we see that domain degradation or even domain back switching may occur in lithium niobate thin films during a thermal annealing process at temperatures on the order of a hundred degrees Celsius. Fortunately, such domain degradation or back switching can be greatly suppressed through a simple pre-heat treatment for the virgin single-domain lithium niobate thin films. It is well known that the domain kinetics in ferroelectric lithium niobate are related to the local field distribution within lithium niobate crystals. At room temperature, the depolarization field, E_d, is fully compensated by the screening field, E_{sceen}, due to surface charges or bulk charges in lithium niobate crystals. When the crystal temperature increases, the spontaneous polarization, P_s, and therefore the depolarization field, E_d, decreases. This breaks the balance between the depolarization field, E_d, and the screening field, E_{sceen}. Therefore, the thermally acted bulk charges, such as protons in lithium niobate may drift in bulk, or the surface charges may accumulated on the surface, to compensate for this field imbalance [28,29]. This will result in a space charge field, E_{sc}, in lithium niobate with its direction antiparallel to the spontaneous polarization, P_s. It is this space charge field that results in the degradation or back switching of the nano-domains in lithium niobate thin films. Note that the component of the space charge field induced by the thermally activated charges are fixed after the crystal is cooled down to the room temperature. This space charge field component induced by the thermally activated charges in crystal is also formed during the pre-heat treatment, and its direction is antiparallel to the spontaneous polarization in the virgin single-domain crystals but parallel to the reversed spontaneous polarization of the stripe or dot domains, which, therefore, results in a great suppression on the degradation of nano-domains. Comprehensive domain kinetics in lithium niobite thin films are an interesting but complicated topic, and they surely deserve a full-length study beyond the scope of this paper; for more details, please refer to Ref. [30].

5. Conclusions

In conclusion, we demonstrated a simple yet effective way to improve the thermal stability of nano-domains fabricated in lithium niobate thin films. We confirmed that the nano-domains in lithium niobate thin films are thermally unstable even at a temperature on the order of ~100 °C, which can be easily reached locally in nano-size photonic structures, due to light absorption. Therefore, such thermal instability of nano-domains could be very detrimental to practical applications, such as PPLN microcavities, PPLN ridge waveguides, and ferroelectric domain memories. We demonstrated that the thermal stability of nano-domains can be greatly improved when the lithium niobate thin film undergoes a pre-heat treatment before the fabrication of nano-domains. This thermal stability improvement is attributed to the generation of a space charge field during the pre-heat treatment, which is parallel to the spontaneous polarization of nano-domains. Our results should be useful for nano-domain-based photonic devices such as PPLN microcavities, PPLN ridge waveguides, and ferroelectric domain memories.

Author Contributions: G.Z. conceived the idea of the work. Y.J. designed and performed the experiments. Z.S. and S.L. participated in the experiments. Y.J. and G.Z. wrote the paper. All authors participated in the data analysis and paper preparation. All authors have read and agreed to the published version of the manuscript.

Acknowledgments: This work was supported by the National Natural Science Foundation of China (NSFC) (11774182, 61475077); the 111 project (B07013); and the Fundamental Research Funds for the Central Universities.

Conflicts of Interest: The authors declare no conflict of interest.

References

1. Turner, E. High-Frequency Electro-Optic Coefficients of Lithium Niobate. *Appl. Phys. Lett.* **1966**, *8*, 303–304. [CrossRef]
2. Kanno, A.; Sakamoto, T.; Chiba, A.; Kawanishi, T.; Higuma, K.; Sudou, M.; Ichikawa, J. 120-Gb/s NRZ-DQPSK signal generation by a thin-lithium-niobate-substrate modulator. *IEICE Electron. Express* **2010**, *7*, 817–822. [CrossRef]
3. Courjal, N.; Benchabane, S.; Dahdah, J.; Ulliac, G.; Gruson, Y.; Laude, V. Acousto-optically tunable lithium niobate photonic crystal. *Appl. Phys. Lett.* **2010**, *96*, 131103. [CrossRef]
4. Hao, Z.; Wang, J.; Ma, S.; Mao, W.; Bo, F.; Gao, F.; Zhang, G.; Xu, J. Sum-frequency generation in on-chip lithium niobate microdisk resonators. *Photonics Res.* **2017**, *5*, 623–628. [CrossRef]
5. Wang, C.; Xiong, X.; Andrade, N.; Venkataraman, V.; Ren, X.F.; Guo, G.C.; Lončar, M. Second harmonic generation in nano-structured thin-film lithium niobate waveguides. *Opt. Express* **2017**, *25*, 6963–6973. [CrossRef] [PubMed]
6. Hu, H.; Yang, J.; Gui, L.; Sohler, W. Lithium niobate-on-insulator (LNOI): status and perspectives. Silicon Photonics and Photonic Integrated Circuits III. *Int. Soc. Opt. Photonics* **2012**, *8431*, 84311D.
7. Poberaj, G.; Hu, H.; Sohler, W.; Guenter, P. Lithium niobate on insulator (LNOI) for micro-photonic devices. *Laser Photonics Rev.* **2012**, *6*, 488–503. [CrossRef]
8. Levy, M.; Osgood, R., Jr.; Liu, R.; Cross, L.; Cargill, G., III; Kumar, A.; Bakhru, H. Fabrication of single-crystal lithium niobate films by crystal ion slicing. *Appl. Phys. Lett.* **1998**, *73*, 2293–2295. [CrossRef]
9. Li, Y.; Wang, C.; Loncar, M. Design of nano-groove photonic crystal cavities in lithium niobate. *Opt. Lett.* **2015**, *40*, 2902–2905. [CrossRef]
10. Wang, J.; Bo, F.; Wan, S.; Li, W.; Gao, F.; Li, J.; Zhang, G.; Xu, J. High-Q lithium niobate microdisk resonators on a chip for efficient electro-optic modulation. *Opt. Express* **2015**, *23*, 23072–23078. [CrossRef]
11. Rabiei, P.; Steier, W.H. Lithium niobate ridge waveguides and modulators fabricated using smart guide. *Appl. Phys. Lett.* **2005**, *86*, 161115. [CrossRef]
12. Weigel, P.O.; Savanier, M.; DeRose, C.T.; Pomerene, A.T.; Starbuck, A.L.; Lentine, A.L.; Stenger, V.; Mookherjea, S. Lightwave circuits in lithium niobate through hybrid waveguides with silicon photonics. *Sci. Rep.* **2016**, *6*, 22301. [CrossRef] [PubMed]
13. Gainutdinov, R.; Volk, T.; Zhang, H. Domain formation and polarization reversal under atomic force microscopy-tip voltages in ion-sliced $LiNbO_3$ films on $SiO_2/LiNbO_3$ substrates. *Appl. Phys. Lett.* **2015**, *107*, 162903. [CrossRef]
14. Hao, Z.; Zhang, L.; Gao, A.; Mao, W.; Lyu, X.; Gao, X.; Bo, F.; Gao, F.; Zhang, G.; Xu, J. Periodically poled lithium niobate whispering gallery mode microcavities on a chip. *Sci. China-Phys. Mech. Astron.* **2018**, *61*, 114211. [CrossRef]

15. Wang, C.; Langrock, C.; Marandi, A.; Jankowski, M.; Zhang, M.; Desiatov, B.; Fejer, M.M.; Loncar, M. Ultrahigh-efficiency wavelength conversion in nanophotonic periodically poled lithium niobate waveguides. *Optica* **2018**, *5*, 1438–1441. [CrossRef]
16. Garcia, V.; Fusil, S.; Bouzehouane, K.; Enouz-Vedrenne, S.; Mathur, N.D.; Barthelemy, A.; Bibes, M. Giant tunnel electroresistance for non-destructive readout of ferroelectric states. *Nature* **2009**, *460*, 81–84. [CrossRef] [PubMed]
17. Guo, R.; You, L.; Zhou, Y.; Lim, Z.S.; Zou, X.; Chen, L.; Ramesh, R.; Wang, J. Non-volatile memory based on the ferroelectric photovoltaic effect. *Nat. Commun.* **2013**, *4*, 1990. [CrossRef]
18. Sharma, P.; Zhang, Q.; Sando, D.; Lei, C.H.; Liu, Y.; Li, J.; Nagarajan, V.; Seidel, J. Nonvolatile ferroelectric domain wall memory. *Sci. Adv.* **2017**, *3*, e1700512. [CrossRef]
19. Jiang, J.; Bai, Z.L.; Chen, Z.H.; He, L.; Zhang, D.W.; Zhang, Q.H.; Shi, J.A.; Park, M.H.; Scott, J.F.; Hwang, C.S.; Jiang, A.Q. Temporary formation of highly conducting domain walls for non-destructive read-out of ferroelectric domain-wall resistance switching memories. *Nat. Mater.* **2018**, *17*, 49–56. [CrossRef]
20. Lindgren, G.; Pena, A.; Zukauskas, A.; Liljestrand, C.; Menaert, B.; Boulanger, B.; Canalias, C. Thermal stability of ferroelectric domain gratings in Rb-doped KTP. *Appl. Phys. Lett.* **2015**, *107*, 082906. [CrossRef]
21. Liu, X.; Kitamura, K.; Terabe, K. Thermal stability of LiTaO$_3$ domains engineered by scanning force microscopy. *Appl. Phys. Lett.* **2006**, *89*, 142906. [CrossRef]
22. Woo, J.; Hong, S.; Min, D.K.; Shin, H.; No, K. Effect of domain structure on thermal stability of nanoscale ferroelectric domains. *Appl. Phys. Lett.* **2002**, *80*, 4000–4002. [CrossRef]
23. Liu, X.Y.; Kitamura, K.; Liu, Y.M.; Ohuchi, F.S.; Li, J.Y. Thermal-induced domain wall motion of tip-inverted micro/nanodomains in near-stoichiometric LiNbO$_3$ crystals. *J. Appl. Phys.* **2011**, *110*, 052009. [CrossRef]
24. Shur, V.Y.; Mingaliev, E.A.; Lebedev, V.A.; Kuznetsov, D.K.; Fursov, D.V. Polarization reversal induced by heating-cooling cycles in MgO doped lithium niobate crystals. *J. Appl. Phys.* **2013**, *113*, 187211. [CrossRef]
25. Shao, G.H.; Bai, Y.H.; Cui, G.X.; Li, C.; Qiu, X.B.; Geng, D.Q.; Wu, D.; Lu, Y.Q. Ferroelectric domain inversion and its stability in lithium niobate thin film on insulator with different thicknesses. *AIP Adv.* **2016**, *6*, 075011. [CrossRef]
26. Saveliev, E.D.; Saveliev, A.P.; Akhmatkhanov, A.R.; Baturin, I.S.; Ya Shur, V. Annealing stability of the domain structure in periodically poled MgO doped lithium niobate single crystals. *Ferroelectrics* **2019**, *542*, 45–51. [CrossRef]
27. Yamada, M.; Saitoh, M. Fabrication of a periodically poled laminar domain structure with a pitch of a few micrometers by applying an external electric field. *J. Appl. Phys.* **1998**, *84*, 2199–2206. [CrossRef]
28. Imbrock, J.; Hanafi, H.; Ayoub, M.; Denz, C. Local domain inversion in MgO-doped lithium niobate by pyroelectric field-assisted femtosecond laser lithography. *Appl. Phys. Lett.* **2018**, *113*, 252901. [CrossRef]
29. Shur, V.Y.; Rumyantsev, E.; Batchko, R.; Miller, G.; Fejer, M.; Byer, R. Domain kinetics in the formation of a periodic domain structure in lithium niobate. *Phys. Solid State* **1999**, *41*, 1681–1687. [CrossRef]
30. Shur, V.Y. Kinetics of ferroelectric domains: Application of general approach to LiNbO$_3$ and LiTaO$_3$. *J. Mater. Sci.* **2006**, *41*, 199–210. [CrossRef]

© 2020 by the authors. Licensee MDPI, Basel, Switzerland. This article is an open access article distributed under the terms and conditions of the Creative Commons Attribution (CC BY) license (http://creativecommons.org/licenses/by/4.0/).

Article

High Homogeneity of Magnesium Doped LiNbO₃ Crystals Grown by Bridgman Method

Xiaodong Yan, Tian Tian *, Menghui Wang, Hui Shen, Ding Zhou, Yan Zhang and Jiayue Xu *

Institute of Crystal Growth, School of Materials Science and Engineering, Shanghai Institute of Technology, Shanghai 201418, China; 176081108@mail.sit.edu.cn (X.Y.); wmh_fzs@163.com (M.W.); hshen@sit.edu.cn (H.S.); dzhou@sit.edu.cn (D.Z.); yanzhang@sit.edu.cn (Y.Z.)

* Correspondence: tiant@sit.edu.cn (T.T.); xujiayue@sit.edu.cn (J.X.); Tel.: +86-021-608-731-17 (T.T.); +86-021-608-734-89 (J.X.); Fax: +86-021-608-731-17 (T.T.); +86-021-608-734-89 (J.X.)

Received: 31 December 2019; Accepted: 25 January 2020; Published: 29 January 2020

Abstract: A series of LiNbO₃ crystals doped with various MgO concentrations (0, 3%, and 5 mol%) was simultaneously grown in one furnace by the modified vertical Bridgman method. The wet chemistry method was used to prepare the polycrystalline powders, and the growth conditions were optimized. The full width at half maximum of high-resolution X-ray rocking curves for (001) reflection of 5 mol% Mg doped lithium niobate (LN) crystal was about 8″, which meant it possessed high crystalline quality. The OH⁻ absorption spectra shifted to 3534.7 cm⁻¹, and the UV absorption edge violet shift indicated that 5 mol% MgO successfully doped in LN and exceeded the threshold. The extraordinary refractive index gradient of 5 mol% Mg doped LN crystal was as small as 2.5×10^{-5}/cm, which exhibited high optical homogeneity.

Keywords: lithium niobate; doping magnesium; Bridgman method; high homogeneity

1. Introduction

Lithium niobate (LiNbO₃, or LN) crystal is one of the most prominent materials for applications in many practical fields, such as optical modulators [1], holographic storage [2], waveguides [3,4], resonators [5], integrated optics devices and three-dimensional (3D) displays, resulting from its superior and diverse physical performance [6,7]. Since the first successful growth by Czochralski method in 1965 [8], crystal growth, photorefractive properties, and theoretical simulations have been studied in depth, and substantial research progress has been reported for LN crystals [9–14]; for example, Ø6″ pure LN crystals with high homogeneity has been reported recently [15].

Normally, LN is a non-stoichiometric compound, and the (Li)/(Nb) ratio of congruent composition is 48.38/51.62 [16,17]. According to the broadly accepted Li-vacancy model, the congruent composition induces a large concentration of intrinsic defects that exist in LN, which mainly are Li vacancies (V_{Li}^-) and anti-site Nb⁵⁺ (Nb_{Li}^{4+}). Small polarons (an electron trapped at Nb_{Li}^{4+}) together with bipolarons (a pair of electrons trapped at adjacent Nb_{Li}^{4+} and Nb_{Nb}^{5+}) play the role of laser-induced optical damage (also named photorefraction) centers in LN [10]. The serious disadvantage of laser-induced optical damage in LN limits its usability in nonlinear optical applications [18]. Doping optical damage resistant additives into LN crystal is an effective approach to suppress the optical damage; especially, Zhong et al. first reported that the laser-induced optical damage could be suppressed by doping MgO with high concentration, exceeding its threshold (about 5 mol%) [19]. The mechanism is when the concentration of MgO exceeded its threshold, Mg²⁺ repelled anti-site Nb⁵⁺ to the site of normal-Nb. Thereby, the formation of small polarons and bipolarons, which serve as an optical damage center, are suppressed remarkably [20]. This discovery impelled magnesium doped LN (LN:Mg) crystals to play a significant role in nonlinear optics and have achieved industrial growth. Consequently, a summarized result of researches on optical grade heavily Mg-doped LN crystals has revealed that most of them have been grown by the Czochralski method with the diameter of Ø1″–2″, and the homogeneity has not been satisfactory [21–25]. Furthermore, only one crystal can be grown in one furnace each time by the Czochralski method, which indicates low growth efficiency.

Bridgman method is one of the main methods for industry crystal growth, such as optical crystals [26], piezoelectric crystals [27,28], ferroelectric crystals [29], and semiconductor crystals [30,31], for its many advantages, especially multiple crystals can be grown in one furnace at the same time, which means high production efficiency. However, there are few reports about the preparation of lithium niobate by the Bridgman method [32,33], especially growing large optical grade heavily Mg-doped LN crystals with high homogeneity is still difficult. Since the lithium niobate crystal belongs to the trigonal crystal system, the crystal is easy to crack. According to previous reports, the growth of crystals becomes more difficult as the concentration of magnesium is increased because the high concentration of MgO in the melt cause (Li)/(Nb) ratio extremely deviate from the congruent composition [34,35]. Besides, the segregation coefficient of MgO deviated from one could induce inhomogeneity along the growth direction [36]. So, the growth process of the Bridgman method for LN:Mg crystals should be optimized.

In this study, we used the Bridgman method to grow Ø2″ LN crystals with different concentrations of magnesium ions. In order to obtain Ø2″ heavily Mg-doped LN with high homogeneity, systematically optimized scheme, including polycrystalline powers preparation, thermal field design, and growth technologies of the Bridgman method, was demonstrated in this work. The homogeneity of LN: Mg crystals was also checked.

2. Materials and Methods

LN:Mg polycrystalline powders were synthesized by a wet chemistry method. $Nb(OH)_5$ (99.99%) was firstly weighted, and excess HCl (38%) was added into a container. Then, they were heated at 90 °C and stirred for 30 min. During this process, the mass of active $Nb_2O_5 \cdot nH_2O$ was precipitated. After being cooled, the observed white precipitate was $NbOCl_3$, which should be completely dissolved by adding deionized water. Malic acid ($C_4H_6O_4$, MA) as the ratio of (MA):(Nb) = 3:1 was added into the former solution and stirred. Then, the pH value of the suspension was adjusted to 8 by the addition of $NH_3 \cdot H_2O$ to get solution A. The chemical reactions happened in producing solution A were as follows:

$$Nb(OH)_5 + 3HCl \rightarrow NbOCl_3 + 4H_2O \quad (1)$$

$$2NbOCl_3 + (n+3)H_2O \rightarrow Nb_2O_5 \cdot nH_2O + 6HCl \quad (2)$$

$$Nb_2O_5 \cdot nH_2O + MA \rightarrow Nb-MA + nH_2O \quad (3)$$

According to the congruent composition of (Li)/(Nb) = 48.38/51.62 and the selected doping concentration of MgO (0, 3 mol%, 5 mol%), Li_2CO_3 (99.99%) and MgO (99.99%) were weighted and dissolved by dilute HCl. Until no more bubbles produced, the pH value was adjusted to 8 by the addition of $NH_3 \cdot H_2O$ to get solution B. Afterward, solutions A and B were mixed to be of high homogeneity by ultrasonic machine. The mixed solution was filtered and spray dried into powders. At last, the powders were sintered at 820 °C for 6 h to obtain LN:Mg polycrystalline powders.

Using the prepared LN:Mg polycrystalline powder, we firstly grew Ø1″ LN:Mg crystals and served them as seed crystals continue to grow Ø2″ LN crystals doped with different MgO concentration of 0, 3%, and 5 mol% by the Bridgman method with multi-crucible [31], which were labeled as LN, LN: Mg3, and LN: Mg5, respectively. The prepared LN: Mg polycrystalline powder was placed in three Pt crucibles with the same dimension of Ø50 mm × 100 mm. After putting them into three Al_2O_3 pipes with the dimension of Ø110 mm × 200 mm, they were simultaneously placed in a furnace. Some mullite fiber mixed with Al_2O_3 powders was used as the thermal insulation material to keep a stable thermal field. In order to be sufficiently melted, the LN: Mg polycrystalline powder was heated by medium-frequency induction and held at 100 °C above the melting point for 2 h. LN, LN: Mg3, and LN: Mg5 crystals were grown in a sealed environment and along the c-axis. In the procedure of crystal growth, the falling rate was governed in the range of 0.5 mm/h to 1.0 mm/h. The vertical temperature gradient above the solid-melt interface was about 0.3 °C/mm, which was measured by using a thermocouple. In order to avoid the cracks occurring in large crystals, they were cooled down to room temperature at a low rate of 30 °C/h after the growth process. Finally, LN: Mg crystals with Ø2" in diameter and 40 mm in length were grown along the c axis. It was necessary to anneal the as-grown crystals at 1230 °C for 30 h to escape thermoelastic stress and improve optical homogeneity. The single-domain structure would be formed by polarization with an electric current density of 7 mA/cm^2 for 20 min at 1190 °C. The 3 mm and 1 mm thick c-oriented plates were cut along the c-axis of the crystals and then polished to optical grade. The distance between the top and bottom part was about 4 cm.

The UV absorption edge wavelength and OH$^-$ spectra of 1-mm-thick plates were measured at room temperature on using a Beckman DU-8B spectrophotometer and Magna-560 Fourier transform IR spectrophotometer, respectively. High-resolution X-ray rocking curves of 3 mm plates were recorded by a Bruker HRXRD-5000 to examine the crystalline quality of LN: Mg3 and LN: Mg5 crystals. The refractive index of 1-mm-thick plates was also measured by METRICON 2010/M prism coupler at 632.8 nm to evaluate the optical homogeneity.

3. Results and Discussions

3.1. Crystal Growth

It is well known that heavily doped LN:Mg crystals remarkably increase the laser damage threshold [37]. However, with the increase in the concentration of MgO, the heavily Mg-doped LN crystals used in optical devices have always been inhomogeneous with low production efficiency using the Czochralski method. Besides, many defects, such as scattering particles and inclusions, have been found in heavily doped LN:Mg crystals. In order to grow Ø2" optical grade heavily doped LN: Mg crystals with high homogeneity, more attention should be paid to its polycrystalline powders preparation, thermal field design, and growth technologies.

Normally, the LN:Mg polycrystalline powders are synthesized through solid reaction, which is mixing Li_2CO_3, Nb_2O_5, and MgO powders and then sintering at about 1100 °C [15]. Though the reaction of Nb_2O_5 with Li_2O, discomposed by Li_2CO_3, could produce LN easily, MgO has very high melt point (2800 °C), weak reaction activity, and low diffusion velocity, which cause difficulties in preparing homogeneous LN:Mg polycrystalline powders and might induce macroscopic inclusions in crystals. Higher reaction temperature or heat preservation with a long time might improve the uniformity of dopant in the polycrystalline powders or melt. However, it is easy to cause the component deviation as the volatilization of Li_2O. Thus, we chose the wet chemistry method with the advantage of low reaction temperature, which was helpful to avoid Li_2O volatilization and enhance the composition homogeneity of LN:Mg. We tested the polycrystalline powders by an X-ray diffractometer, and the results are shown in Figure 1. Based on the PDF#74–2238, the XRD spectrum showed that the diffraction peaks and relative intensity of LN, LN: Mg3, and LN: Mg5 polycrystalline materials were very similar to those of lithium niobate without any obvious peak shift or second phase. That indicated that LN could be successfully prepared by the wet chemical method, and doping Mg^{2+} had a negligible effect on diffraction data. Good quality of LN:Mg polycrystalline powders laid the foundation for crystal growth. Besides, most of doped LN crystals were grown by using pure LN crystals as seeds, but we used the Bridgman method to grow Ø1" LN crystals with different concentrations of MgO and then cut as seeds for growing Ø2" LN:Mg crystals.

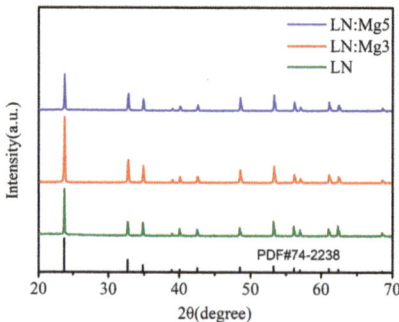

Figure 1. XRD results of LN:Mg polycrystalline powders prepared by the wet chemical method. LN—lithium niobate.

A stable and suitable thermal field is crucial to growing Ø2" LN: Mg crystals with high quality. In the past, only Al_2O_3 powders were usually selected as insulation materials, but the heating time was long due to the low thermal conductivity of Al_2O_3 powders. In order to avoid the shortcomings, some mullite fiber was mixed with Al_2O_3 powders and served as insulation materials. About 35% of energy saving could be realized because of the higher thermal conductivity of mullite fiber compared to Al_2O_3 powders, which was also beneficial in keeping high stability of the thermal field. Moreover, with the excellent thermal insulation of mullite fiber mixed with Al_2O_3 powders, a small radial temperature gradient was obtained to avoid cracks caused by large thermal stress. The vertical temperature gradient $\left(\frac{\partial T}{\partial Z}\right)_s$ above the solid-melt interface was designed as 1 °C/mm, which was smaller than other LN crystals with a small diameter [27]. The reason is that the vertical temperature gradient, as seen from Equation (4), should be controlled in a certain range and has an inverse relationship with the diameter of the crystal [29].

$$\frac{2\varepsilon_b}{\alpha R^{3/2}} \bullet \left(\frac{2}{h}\right)^{1/2} \geq \left(\frac{\partial T}{\partial Z}\right)_s \geq \{-\frac{k_l m V_T (C_l(B)(1-k^*))}{D[k^* + (1-k^*)\exp(v_T \delta/D)]} + L\rho v_T\}/k_s \qquad (4)$$

where α_a, ε_b, α, R, h, k_s, and v_T are the a-direction thermal expansion coefficient, fracture strain, thermal expansion coefficient, diameter, heat exchange coefficient, thermal conductivity, and falling rate (growth velocity of crystal), respectively; k_l and $C_l(B)$ are the thermal conductivity and bulk concentration of melt; m, δ, D, k^*, L, and ρ are the liquidus slope, depth of solute boundary layer, diffusion coefficient, segregation coefficient at the interface, crystalline latent heat and density, respectively. Besides, the falling rate was optimizing from 1.5 mm/h–2.0 mm/h to 0.5 mm/h–0.8 mm/h at different growth stages. In the crystal growth experiment, we set the descending speed to 0.5 mm/h and extended the holding time to 8 h, which was more conducive to Mg^{2+} entering the crystal lattice. The lower falling rate could provide enough time for the sufficient diffusion of Mg^{2+} at the solid-melt interface and improving Mg^{2+} distribution homogeneity in the crystal. The flat or slight convex shape was helpful to Mg^{2+} diffused along the vertical and parallel direction of the solid-melt interface.

Based on the above improvements, colorless, transparent, crack-free, and inclusions free LN:Mg single crystals were grown. The cut and polished LN: Mg5 crystal with a length of 4 cm is shown in Figure 2. High-resolution X-ray rocking curves are widely used in checking the crystalline quality of single crystals. The narrower full width at half maximum (FWHM) of single crystals means higher crystalline quality. Here, the X-ray rocking curves for c-plates of LN:Mg3 crystal and LN:Mg5 crystal are given in Figure 3. The FWHM was measured to be 8" and 14" for (001) reflection of LN:Mg5 and LN:Mg3 crystal, respectively, which was better than the reported results [38]. It implied that they possessed high structural quality with few dislocations and thermal stress [39]. This proved that the Bridgman method could grow LN:Mg crystals with higher crystallinity.

Figure 2. Cut and polished LN: Mg5 crystal grew by the Bridgman method.

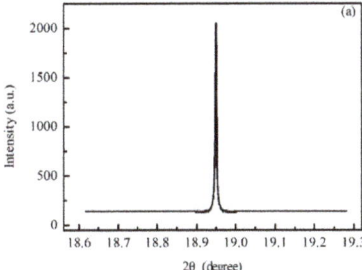 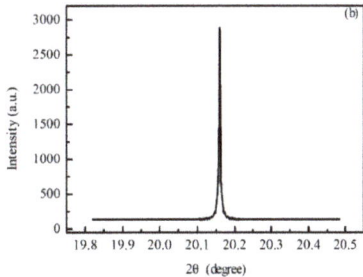

Figure 3. X-ray rocking curves of (001) reflection in LN:Mg crystals. (**a**) and (**b**) are for LN:Mg3 crystal and LN:Mg5 crystal, respectively.

3.2. Characterization

As we know, hydrogen ion can be introduced into LN crystals by means of water vapor during the growth of the crystals and forms as OH⁻ exists in the lattice. As OH⁻ is heavily sensitive to the surrounding environment, OH⁻ spectra are usually used to investigate the composition and defect structures of LN. An infrared absorption band near 2.87 μm (~3480 cm^{-1}) in pure LN crystal was first reported by Smith et al. [40]. Herrington et al. demonstrated that the absorption band was caused by the stretching vibrations of OH⁻ ions [41]. For doped LN crystals, it is well known that when the optical damage resistant dopants, such as Mg^{2+}, In^{3+}, and Hf^{4+}, are doped with the concentration exceeding their threshold, the OH⁻ absorption band shifts from the position at 3484 cm^{-1} of pure LN to higher wavenumbers [42,43]. As shown in Figure 4, LN and LN:Mg3 crystals showed a broad OH⁻ absorption band peak at approximately 3484 cm^{-1}, while the OH⁻ peak of LN:Mg5 crystal shifted to the higher wavenumber of 3534.7 cm^{-1}. It was proposed that in LN:Mg crystals, Mg^{2+} ions occupying Li-sites would push the Nb_{Li}^{4+} ions to the normal Nb-sites until all of the Nb_{Li}^{4+} were clean up when Mg concentration reached the threshold. Above the concentration threshold of Mg, additional Mg^{2+} ions would occupy Nb-sites. The position of 3534.7 cm^{-1} nearly coincided with the result of [42] and related to the OH⁻ vibration formation in (Mg_{Nb}^{2+}- OH⁻) complex. It indicated that MgO was effectively doped into LN crystals, and 5 mol% had exceeded the threshold. Besides, as OH⁻ absorption band peaks of different positions in different plates or the same plate were nearly centered at the same wavenumber, it reflected that MgO distributed homogenously in LN:Mg5 crystal.

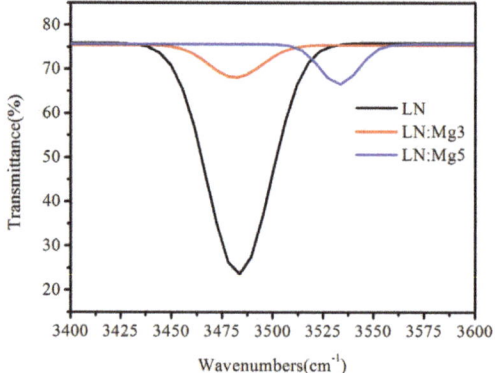

Figure 4. The OH⁻ spectroscopy of LN:Mg crystals.

Besides, the UV absorption edge of LN is also sensitive to defects [44]. Figure 5a shows that the UV absorption edge of LN:Mg3 and LN:Mg5 crystals was attributed to short wavelength compared to LN, especially the violet shift could be seen more obviously for LN:Mg5. The UV absorption edge of LN crystals has also been attributed to the presence of Li vacancies, actually of O^{2-} ions in the vicinity of $V_{Li}{}^-$, forming as the defect of ($V_{Li}{}^- - O^{2-}$) [45]. As mentioned above, for LN:Mg crystals, Mg^{2+} ions pushed the $Nb_{Li}{}^{4+}$ ions to the normal Nb-sites and formed $Mg_{Li}{}^{1+}$, until all of the $Nb_{Li}{}^{4+}$ dismissed when the MgO concentration exceeded the threshold. Compared with $Nb_{Li}{}^{4+}$ that needs four cationic vacancies $V_{Li}{}^-$ for keeping the electric charge equilibrium [46], $Mg_{Li}{}^{1+}$ only needs one $V_{Li}{}^-$ for charge compensation. Thereby, the decrement of the ($V_{Li}{}^- - O^{2-}$) defect concentration caused the observation of the UV absorption edge violet shift with the increment in MgO doping concentration. Especially, the $Nb_{Li}{}^{4+}$ dismissal induced a more obvious violet shift in LN:Mg5 because of the MgO concentration exceeding the threshold. This result was also in accordance with the results of OH⁻ spectra. We also compared the transmittance of the top and bottom parts for LN:Mg5, as shown in Figure 5b. It was clear that the two curves almost coincided throughout the 4 cm long crystal, indicating the crystal possessed a nice uniformity.

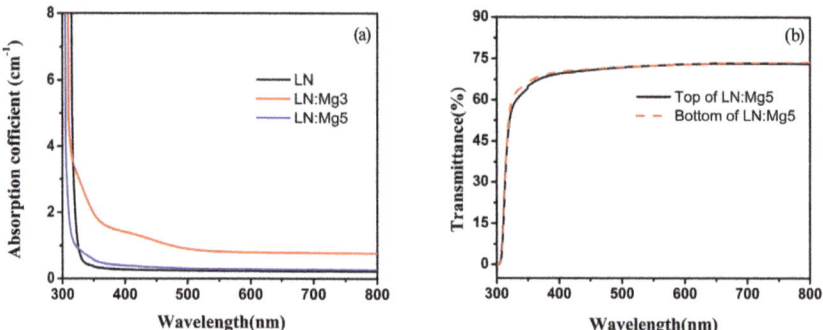

Figure 5. (a) is the UV absorption edge of LN:Mg crystals and (b) is the transmittance of the top and bottom plate in LN:Mg5 crystal.

High optical homogeneity is significant for the application of the nonlinear optical crystal. According to the high compositional homogeneity discussed above, LN:Mg5 crystal should also have high optical homogeneity. For the extraordinary refractive index n_e is sensitive to the composition while the ordinary refractive index n_o is not, the gradient of the extraordinary refractive index $\delta\Delta n_e$ was measured to examine the optical homogeneity of LN:Mg5 crystal. As listed in Table 1, the difference between the average extraordinary refractive index $\delta\Delta n_e$ of the top and bottom plate in LN:Mg5 crystal was 1×10^{-4}. Since the distance of the two plates was about 4 cm, the gradient of the extraordinary refractive index $\delta\Delta n_e$ was about 2.5×10^{-5}/cm, which exhibited high optical homogeneity of the crystal. The optical homogeneity was about two times higher than the reported high optical homogeneous LN:Mg crystal, of which $\delta\Delta n_e$ was about 5.11×10^{-5}/cm [47].

Table 1. The extraordinary refractive index of the top and bottom plate in lithium niobate (LN):Mg5 crystal.

Samples	Extraordinary Refractive Index (n_e)					
	Position 1	Position 2	Position 3	Position 4	Position 5	Average
Bottom of LN:Mg5	2.1919	2.1917	2.1917	2.1918	2.1917	2.19176
Top of LN:Mg5	2.1916	2.1917	2.1918	2.1917	2.1915	2.19166

4. Conclusions

LN: Mg single crystals doped with different MgO concentrations were grown successfully in one furnace at the same time by the Bridgman method. Wet chemistry method was employed for LN: Mg polycrystalline powders preparation to enhance doping homogeneity of MgO and avoid Li_2O volatilization. The critical growth conditions included a small vertical temperature gradient, a low growth rate, adjusting the descending speed at different growth stages to keep a micro convex solid-melt interface. X-ray rocking curves of (001) reflection showed that the LN: Mg5 crystal had a high crystallinity. Compared with the congruent LN crystal, the OH^- absorption peaks and the ultraviolet absorption edge exhibited that the doping concentration of MgO exceeded the threshold. Moreover, LN: Mg5 crystal had high optical homogeneity for the extraordinary refractive index gradient that was as small as 2.5×10^{-5}/cm.

Author Contributions: T.T. conceived and designed the experiments. T.T., X.Y. and M.W. performed the experiments, analyzed the data. H.S., D.Z., and Y.Z. contributed the measurements. J.X. contributed useful and deep discussions. T.T. and X.Y. wrote the manuscript. All authors read and approved the final version of the manuscript to be submitted.

Funding: This work was partially supported by the National Natural Science Foundation of China (61605116, 51972208, and 51972213), Jiangsu Planned Projects for Postdoctoral Research Funds (1501131C).

Conflicts of Interest: The authors declare no conflict of interest.

References

1. Wang, C.; Zhang, M.; Stern, B.; Lipson, M.; Loncar, M. Nanophotonic lithium niobate electro-optic modulators. *Opt. Express* **2018**, *26*, 1547–1555. [CrossRef] [PubMed]
2. Dhar, L.; Curtis, K.; Fäcke, T. Holographic data storage: Coming of age. *Nat. Photonics* **2008**, *2*, 403–405. [CrossRef]
3. Wang, Y.; Zhou, S.X.; He, D.H.; Hu, Y.; Chen, H.X.; Liang, W.G.; Yu, J.H.; Guan, H.Y.; Luo, Y.H.; Zhang, J.; et al. Electro-optic beam deflection based on a lithium niobate waveguide with microstructured serrated electrodes. *Opt. Lett.* **2016**, *41*, 4739–4742. [CrossRef] [PubMed]
4. Bazzan, M.; Sada, C. Optical waveguides in lithium niobate: Recent developments and applications. *Appl. Phys. Rev.* **2015**, *2*, 040603. [CrossRef]
5. Jiang, H.W.; Luo, R.; Liang, H.X.; Chen, X.F.; Chen, Y.P.; Lin, Q. Fast response of photorefraction in lithium niobate microresonators. *Opt. Lett.* **2017**, *42*, 3267–3270. [CrossRef]
6. Tu, D.; Xu, C.N.; Yoshida, A.; Fujihala, M.; Hirotsu, J.; Zheng, X.G. $LiNbO_3:Pr^{3+}$: A multipiezo material with simultaneous piezoelectricity and sensitive piezoluminescence. *Adv. Mater.* **2017**, *29*, 1606914. [CrossRef]
7. Gopalan, K.K.; Janner, D.; Nanot, S.; Parret, R.; Lundeberg, M.B.; Koppens, F.H.L.; Pruneri, V. Mid-infrared pyroresistive graphene detector on $LiNbO_3$. *Adv. Opt. Mater.* **2017**, *5*, 1600723. [CrossRef]

8. Ballman, A.A. Growth of piezoelectric and ferroelectric materials by the Czochralski technique. *J. Am. Ceram. Sot.* **1965**, *48*, 112–113. [CrossRef]
9. Lengyel, K.; Péter, Á.; Kovács, L.; Corradi, G.; Pálfalvi, L.; Hebling, J.; Unferdorben, M.; Dravecz, G.; Hajdara, I.; Szaller, Z. Growth, defect structure, and THz application of stoichiometric lithium niobate. *Appl. Phys. Rev.* **2015**, *2*, 040601. [CrossRef]
10. Hesselink, L.; Orlov, S.S.; Liu, A.; Akella, A.; Lande, D.; Neurgaonkar, R.R. Photorefractive materials for nonvolatile volume holographic data storage. *Science* **1998**, *282*, 1089–1094. [CrossRef]
11. Schmidt, W.G.; Albrecht, M.; Wippermann, S.; Blankenburg, S.; Rauls, E.; Fuchs, F.; Rödl, C.; Furthmüller, J.; Hermann, A. $LiNbO_3$ ground- and excited-state properties from first-principles calculations. *Phys. Rev. B* **2008**, *77*, 035106. [CrossRef]
12. Bernert, C.; Hoppe, R.; Wittwer, F.; Woike, T.; Schroer, C.G. Ptychographic analysis of the photorefractive effect in $LiNbO_3$:Fe. *Opt. Express* **2017**, *25*, 31640–31650. [CrossRef] [PubMed]
13. Yang, Y.P.; Buse, K.; Psaltis, D. Photorefractive recording in $LiNbO_3$:Mn. *Opt. Lett.* **2002**, *27*, 158–160. [CrossRef] [PubMed]
14. Ren, L.Y.; Liu, L.R.; Liu, D.A.; Zu, J.F.; Luan, Z. Optimal switching from recording to fixing for high diffraction from a $LiNbO_3$:Ce:Cu photorefractive nonvolatile hologram. *Opt. Lett.* **2004**, *29*, 186–188. [CrossRef]
15. Wang, S.; Ji, C.; Dai, P.; Shen, L.; Bao, N. The growth and characterization of six inch lithium niobate crystal with high homogeneity. *Cryst. Eng. Comm.* **2020**. [CrossRef]
16. Byer, R.L.; Young, J.F.; Feigelson, R.S. Growth of high-quality $LiNbO_3$ crystals from the congruent melt. *J. Appl. Phys.* **1970**, *41*, 2320–2325. [CrossRef]
17. O'BRYAN, H.M.; Gallagher, P.K.; Brandle, C.D. Congruent composition and Li-rich phase boundary of $LiNbO_3$. *J. Am. Ceram. Soc.* **1985**, *68*, 493–496. [CrossRef]
18. Ashkin, A.; Boyd, G.D.; Dziedzic, J.M.; Smith, R.G.; Ballman, A.A.; Levinstein, J.J.; Nassau, K. Optically-induced refractive index inhomogeneities in $LiNbO_3$ and $LiTaO_3$. *Appl. Phys. Lett.* **1966**, *9*, 72–74. [CrossRef]
19. Zhong, G.G.; Jian, J.; Wu, Z. Measurement of optically induced refractive-index damage of lithium niobate doped with different concentrations of MgO. *J. Opt. Soc. Am.* **1980**, *70*, 631.
20. Liu, J.J.; Zhang, W.L.; Zhang, G.Y. Defect chemistry analysis of the defect structure in Mg-doped $LiNbO_3$ crystals. *Phys. Stat. Sol(A).* **1996**, *156*, 285–291. [CrossRef]
21. Yao, L.F.; Li, J.L.; Liu, J.H.J. The study of growth on magnesium doped lithium niobate crystals. *J. Changchun Inst. Opt. Fine Mech.* **1994**, *17*, 65–68.
22. Kim, I.W.; Park, B.C.; Jin, B.M.; Bhalla, A.S.; Kim, J.W. Characteristics of MgO-doped $LiNbO_3$ crystals. *Mater. Lett.* **1995**, *24*, 157–160. [CrossRef]
23. Bae, S.I.; Ichikawa, J.; Shimamura, K.; Onodera, H.; Fukuda, T. Doping effects of Mg and/or Fe ions on congruent $LiNbO_3$ single crystal growth. *J. Cryst. Growth.* **1997**, *180*, 94–100. [CrossRef]
24. Chen, Y.L.; Guo, J.; Lou, C.B.; Yuan, J.W.; Zhang, W.L.; Chen, S.L.; Zhang, G.Y. Crystal growth and characteristics of 6.5 mol% MgO-doped $LiNbO_3$. *J. Cryst. Growth.* **2004**, *263*, 427–430. [CrossRef]
25. Palatnikov, M.N.; Birukova, I.V.; Masloboeva, S.M.; Makarova, O.V.; Manukovskaya, D.V.; Sidorov, N.V. The search of homogeneity of $LiNbO_3$ crystals grown of charge with different genesis. *J. Cryst. Growth.* **2014**, *386*, 113–118. [CrossRef]
26. Yoshimura, M.; Sakata, S.I.; Iba, H.; Kawano, T.; Hoshikawa, K. Vertical Bridgman growth of Al_2O_3/YAG: Ce melt growth composite. *J. Cryst. Growth.* **2015**, *416*, 100–105. [CrossRef]
27. Shi-Ji, F.; Guan-Shun, S.; Wen, W.; Jin-Long, L.; Xiu-hang, L. Bridgman growth of $Li_2B_4O_7$ crystals. *J. Cryst. Growth.* **1990**, *99*, 811–814. [CrossRef]
28. Nishimura, E.; Okano, K.; Iida, J.; Hoshikawa, K. $LiTaO_3$ Single Crystal Growth by the Vertical Bridgman Technique. *Crystl. Res. Technol.* **2018**, *53*, 1800044. [CrossRef]
29. Liu, W.B.; Tian, T.; Yang, B.B.; Xu, J.Y.; Liu, H.D. Bridgman growth and luminescence properties of dysprosium doped lead potassium niobate crystal. *J. Cryst. Growth.* **2017**, *468*, 462–464. [CrossRef]
30. Jin, M.; Lin, S.; Li, W.; Chen, Z.; Li, R.; Wang, X.; Pei, Y. Fabrication and thermoelectric properties of single-crystal argyrodite Ag_8SnSe_6. *Chem. Mater.* **2019**, *31*, 2603–2610. [CrossRef]
31. Jin, M.; Shen, H.; Fan, S.J.; He, Q.B.; Xu, J.Y. Industrial growth and characterization of Si-doped GaAs crystal by a novel multi-crucible Bridgman method. *Cryst. Res. Technol.* **2017**, *52*, 1700052. [CrossRef]

32. Chen, H.; Xia, H.; Wang, J.; Zhang, J.; Xu, J.; Fan, S. Growth of LiNbO$_3$ crystals by the Bridgman method. *J. Cryst. Growth.* **2003**, *256*, 219–222. [CrossRef]
33. Xu, X.; Liang, X.; Li, M.; Solanki, S.; Chong, T.C. Two-color nonvolatile holographic recording in Bridgman-grown Ru: LiNbO$_3$ crystals. *J. Cryst. Growth.* **2008**, *310*, 1976–1980. [CrossRef]
34. Iyi, N.; Kitamura, K.; Yajima, Y.; Kimura, S.; Furukawa, Y.; Sato, M. Defect structure model of MgO-doped LiNbO$_3$. *J. Solid. State. Chem.* **1995**, *118*, 148–152. [CrossRef]
35. Niwa, K.; Furukawa, Y.; Takekawa, S.; Kitamura, K. Growth and characterization of MgO doped near stoichiometric LiNbO$_3$ crystals as a new nonlinear optical material. *J. Cryst. Growth.* **2000**, *208*, 493–500. [CrossRef]
36. Furukawa, Y.; Sato, M.; Nitanda, F.; Ito, K. Growth and characterization of MgO-doped LiNbO$_3$ for electro-optic devices. *J. Cryst. Growth.* **1990**, *99*, 832–836. [CrossRef]
37. Bryan, D.A.; Gerson, R.; Tomaschke, H.E. Increased optical damage resistance in lithium niobate. *Appl. Phy. Lett.* **1984**, *44*, 847–849. [CrossRef]
38. Bhatt, R.; Bhaumik, I.; Ganesamoorthy, S.; Bright, R.; Soharab, M.; Karnal, A.K.; Gupta, P.K. Control of intrinsic defects in lithium niobate single crystal for optoelectronic applications. *Crystals.* **2017**, *7*, 23. [CrossRef]
39. Yao, S.H.; Hu, X.B.; Wang, J.Y.; Liu, H.; Chen, X.F. Growth and characterization of near stoichiometric LiNbO$_3$ single crystal. *Cryst. Res. Technol.* **2007**, *42*, 114–118. [CrossRef]
40. Smith, R.G.; Fraser, D.B.; Denton, R.T.; Rich, T.C. Correlation of reduction in optically induced refractive index inhomogeneity with OH content in LiTaO$_3$ and LiNbO$_3$. *J. Appl. Phys.* **1968**, *39*, 4600–4602. [CrossRef]
41. Herrington, J.R.; Dischler, B.; Räuber, A.; Schneider, J. An optical study of the stretching absorption band near 3 microns from OH$^-$ defects in LiNbO$_3$. *Solid State Commun.* **1973**, *12*, 351–354. [CrossRef]
42. Kong, Y.F.; Deng, J.C.; Zhang, W.L.; Wen, J.K.; Zhang, G.Y.; Wang, H.F. OH$^-$ absorption spectra in doped lithium niobate crystals. *Phys. Lett. A* **1994**, *196*, 128–132. [CrossRef]
43. Kokanyan, E.P.; Razzari, L.; Cristiani, I.; Degiorgio, V.; Gruber, J.B. Reduced photorefraction in hafnium-doped single-domain and periodically poled lithium niobate crystals. *Appl. Phys. Lett.* **2004**, *4*, 1880–1882. [CrossRef]
44. Földvári, I.; Polgár, K.; Voszka, R.; Balasanyan, R.N. A simple method to determine the real composition of LiNbO$_3$ crystals. *Cryst. Res. Technol.* **1984**, *19*, 1659–1661. [CrossRef]
45. Li, X.; Kong, Y.; Liu, H.; Sun, L.; Xu, J.; Chen, S.; Zhang, G. Origin of the generally defined absorption edge of non-stoichiometric lithium niobate crystals. *Solid State Commun.* **2007**, *141*, 113–116. [CrossRef]
46. Kityk, I.V.; Makowska-Janusik, M.; Fontana, M.D.; Aillerie, M.; Abdi, F. Nonstoichiometric defects and optical properties in LiNbO$_3$. *J. Phys. Chem. B.* **2001**, *105*, 12242–12248. [CrossRef]
47. Ferriol, M.; Dakki, A.; Cohen-Adad, M.T.; Foulon, G.; Boulon, G. Growth and characterization of mgo-doped single-crystal fibers of lithium niobate in relation to high temperature phase equilibria in the ternary system Li$_2$O-Nb$_2$O$_5$-MgO. *J. Cryst. Growth.* **1997**, *178*, 529–538. [CrossRef]

© 2020 by the authors. Licensee MDPI, Basel, Switzerland. This article is an open access article distributed under the terms and conditions of the Creative Commons Attribution (CC BY) license (http://creativecommons.org/licenses/by/4.0/).

Article

Design and Optimization of Proton Exchanged Integrated Electro-Optic Modulators in X-Cut Lithium Niobate Thin Film

Huangpu Han [1,2], Bingxi Xiang [3,*], Tao Lin [3], Guangyue Chai [3] and Shuangchen Ruan [3]

1. College of Electric and Electronic Engineering, Zibo Vocational Institute, Zibo 255314, China; pupuhan@126.com
2. School of Physics, Shandong University, Jinan 250100, China
3. College of New Materials and New Energies, Shenzhen Technology University, Shenzhen 518118, China; lintao@sztu.edu.cn (T.L.); chaiguangyue@sztu.edu.cn (G.C.); ruanshuangchen@sztu.edu.cn (S.R.)
* Correspondence: xiangbingxi@sztu.edu.cn; Tel.: +86-135-3085-0236

Received: 29 September 2019; Accepted: 22 October 2019; Published: 24 October 2019

Abstract: In this study, we designed, simulated, and optimized proton exchanged integrated Mach-Zehnder modulators in a 0.5-µm-thick x-cut lithium niobate thin film. The single-mode conditions, the mode distributions, and the optical power distribution of the lithium niobate channel waveguides are discussed and compared in this study. The design parameters of the Y-branch and the separation distances between the electrodes were optimized. The relationship between the half-wave voltage length production of the electro-optic modulators and the thickness of the proton exchanged region was studied.

Keywords: electro-optic modulator; lithium niobate thin film; proton exchange; Mach-Zehnder; integrated optics devices

1. Introduction

Electro-Optic (E-O) modulators have recently attracted growing attention in ultra-compact photonic integrated circuits (PICs) [1]. They have extensive applications in optical telecommunication networks and microwave-photonic systems [2]. The Mach-Zehnder interferometer (M-ZI) is one of the most important interference structures in modulators because of its simple design and manufacture, with the existence of a reference arm that compensates for the common-mode effect [3]. Many types of M-ZI-based applications for optical communication have been investigated, such as switches/modulators [4,5], multi/demultiplexers [6,7], and splitters [8,9].

Lithium niobate (LiNbO$_3$, LN) is one of the most remarkable optical crystal materials due to its combination of excellent E-O and nonlinear optical characteristics [10]. Due to the high E-O coefficient (r$_{33}$ = 31.2 pm/V) in LN, high-quality E-O modulators of this type are very valuable in optical communication [11–15]. In the last decade, high-refractive-index contrast in the form of lithium niobate thin film bonded to a SiO$_2$ layer (lithium niobate on insulator, LNOI) has emerged as an ideal platform for integrated high-performance modulators [16–19]. A basic challenge in the production of M–ZI modulators in LNOI is the fabrication of high-quality waveguide structures. A few techniques have been developed for fabricating waveguides in LN, including dry-etching [20–22], proton exchange (PE) [23], and chemo-mechanical polishing [24]. Compared with other methods, PE is low-cost, has low propagation loss, and is a mature manufacturing method that is compatible with the LN optical waveguide industry [25,26]. Compared with rough-etched side walls, PE waveguides have smooth boundaries. However, to the best of the authors' knowledge, to date there have been few reports on proton-exchanged electro-optic modulators in LNOI [23].

In this research, we simulated and analyzed a proton-exchanged E-O M-ZI modulator in an x-cut LNOI. Based on the full-vectorial finite-difference method [27], the single-mode conditions of the PE waveguides were investigated, the bending losses of the Y-branch structures were analyzed, and the propagation losses of the PE waveguides with different separation distances between electrodes were simulated. The half-wave voltages of the devices were calculated using the finite difference beam propagation method (FD-BPM) [28,29]. The optimized half-wave voltage-length product ($V_\pi \cdot L$) was approximately 10.2 V·cm.

2. Device Design and Methods

The material of the device studied was an x-cut LN thin film bonded to a SiO_2 layer deposited on an LN-substrate [17]. The thicknesses of the LN thin film and the SiO_2 layer were 0.5 µm and 2 µm [30], respectively. The structures were cladded with 2-µm-thick SiO_2 layers after the PE waveguides and electrodes were fabricated. Figure 1a shows a schematic of the M-ZI. The input wave was emitted into a directional coupler. The input power was divided equally into the two output waveguides with a first directional coupler. The two waveguides formed the two arms of the M-ZI. On both arms, opposite electric fields were applied to modify the refractive of the LN and thus change the phase of the wave propagating through that arm. The two waves were then combined into another 50/50 directional coupler. By varying the applied voltage, the amount of light emitted from the two output waveguides could be continuously controlled.

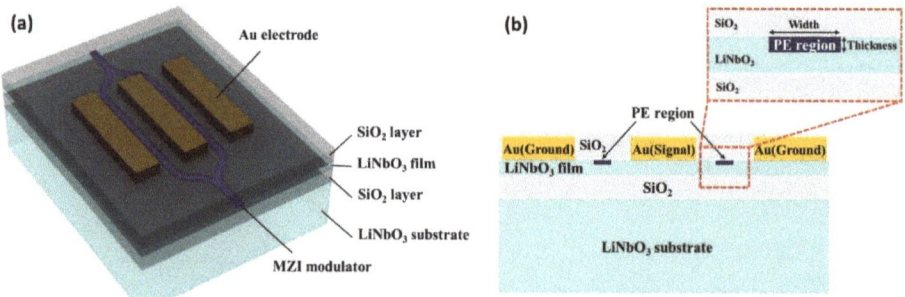

Figure 1. (a) Schematic of the M-ZI. (b) The cross-sectional schematic of the M-ZI and the channel waveguides.

Figure 1b shows the schematic cross-sections of the M-ZI and the channel waveguide. The lateral diffusion could be neglected when the thickness of the PE was much lower than the mask width, and a rectangular step-like refractive index profile could be formed during the PE process, as shown in the inset of Figure 1b. The PE region formed stripe-loaded channel waveguides, and the LN thin film on both sides of the PE region formed planar waveguides. The PE only increased the extraordinary refractive index (n_e) of the LN crystal and it therefore supported only one type of mode in the channel waveguides (the transverse electric (TE) mode in the x-cut LNOI). The ordinary and extraordinary refractive index changes were −0.05 and 0.08, respectively. Table 1 shows the refractive indices of the material at the wavelength of 1.55 µm. In previous studies on bulk LN, the PE waveguides generally suffered from a dramatically reduced E-O coefficient for the electro-optic devices [31]. Since the E-O coefficient of the unannealed proton exchange region was close to zero, it was set to zero in the simulation.

Table 1. Refractive indices of the material in simulation (λ = 1.55 μm).

Material	n_o	n_e
LN [32]	2.211	2.138
PE [23]	2.161	2.218
SiO$_2$	1.46	1.46

The full-vectorial finite difference method was regarded as a simple and effective method to solve the bending loss and the mode distribution. The finite difference algorithm was used to mesh the geometry of the waveguide. This algorithm had the ability to adapt to the arbitrary waveguide structure. After the structure was meshed, Maxwell's equations were then transformed into matrix eigenvalue problems and solved by sparse matrix techniques to obtain the effective indices and the mode distribution of the waveguide modes [33]. The FD-BPM was employed to simulate the M-ZI. The method consisted of marching the input optical field over small distances in the dielectric media with the use of a fast Fourier transform. In each propagational step, the plane spectrum was used to simulate the optical field in the spectral domain, and due to the medium inhomogeneity, a phase correction was introduced in the spatial domain [34].

3. Results and Discussion

The single-mode conditions were simulated to prevent the distortion of the signal during transmission. We calculated the modal curves of the PE waveguides at a wavelength of λ = 1.55 μm. The effective indices of the TE mode in the PE waveguides as a function of the width for different thicknesses of the PE region are presented in Figure 2a. As the width and thickness of the PE waveguides decreased, the effective refractive index decreased, and the more high-order modes disappeared. The TE$_0$ modes represented the fundamental TE modes. The cutoff dimension of the PE waveguide for the TE mode between the single- and multi-mode conditions was calculated, as shown in Figure 2b. Any dimensions beneath the curves fulfilled the single-mode condition. As the cut-off width increased, the PE thickness decreased.

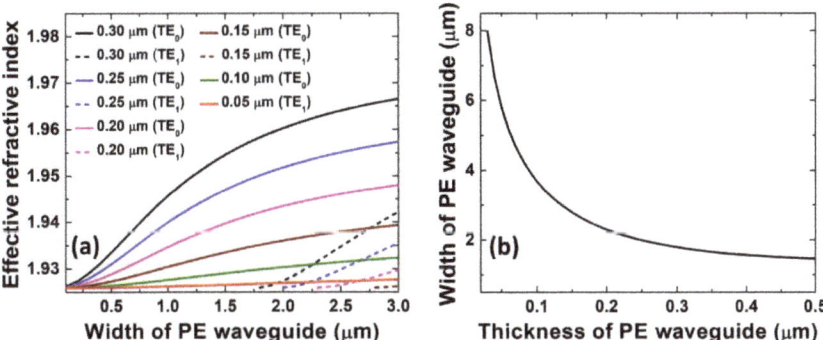

Figure 2. (a) Effective indices of the transverse electric (TE) modes in the proton exchange (PE) waveguides as a function of the width for different thicknesses of the PE waveguides. (b) Cut-off dimensions of the PE region for the TE mode.

A small mode size enabled the development of ultra-compact PICs and strengthened the E-O effect. For the channel waveguide in the LNOI, due to the large refractive index contrast between the LN layer and the SiO$_2$ cladding, the light was strongly confined, resulting in a smaller mode size. Figure 3a shows the simulation results of the relationship between the mode size (the 1/e intensity in the vertical and horizontal directions formed the two axes of the ellipse) and the width and thickness of the PE waveguides. The mode size decreased with the increasing PE thickness. When the width initially increased, the confinement of the light became strong, which also led to a smaller mode size. As the waveguide width increased further, the PE region expanded and the mode size became larger. The shape of the smallest mode size decreased in width as the thickness increased. For the composite strip waveguide, the optical power was mainly divided into three parts. The first part was in the PE region, the second part was in the LN layer without PE, and the third part was in the SiO_2 cladding layer. The E-O effect ascended with the increasing optical power in the LN layer without PE. This required most of the optical power to be concentrated in the LN layer without PE. Figure 3b shows the optical powers in the LN layer without PE and the PE region for the TE modes. The optical power in the LN layer without PE increased slightly with the shrinking of the PE width and thickness. Therefore, the width and thickness of the PE strip had a certain influence on the optical power distribution. Considering the single-mode conditions, mode sizes, and optical power distribution, the widths of the PE waveguides were all selected as 1.2 μm in the following simulations.

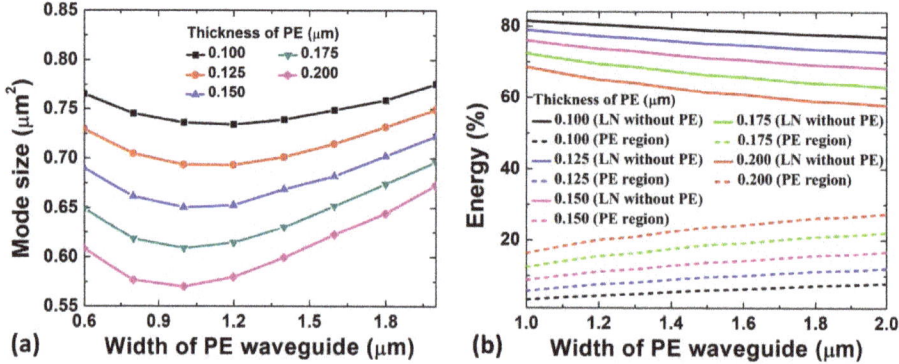

Figure 3. (a) Relationship between the mode size and the width and thickness of the PE waveguide. (b) Relationship between the optical power and the width and thickness of the PE waveguide.

Figure 4a shows the schematic of the Y-branch. It consisted of two symmetrical arms between one input and two output straight waveguides. Each arm consisted of two identical circular arcs of radius R which had the same width as the input and output waveguides. Since the structure involved a bend waveguide (circular arcs on each arm), the relationship between the bending loss and the bending radius was as shown in Figure 4b. The bending loss increased sharply with the decreasing bending radius and thickness of the PE region.

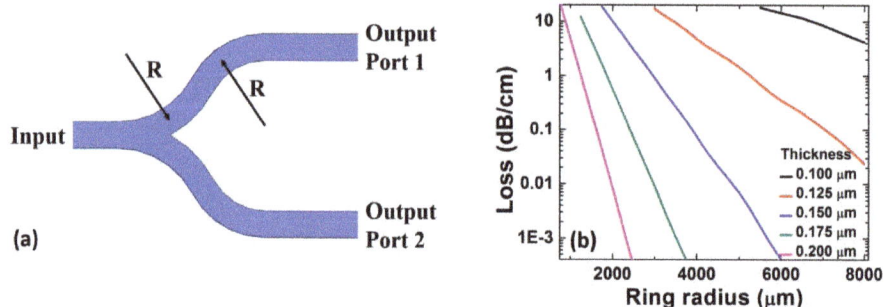

Figure 4. (a) Schematic of the symmetrical Y-Branch. (b) The dependence of the bending loss on the bending radius of the channel waveguide with a 1.2 μm width, using the thickness of the PE waveguide as the parameter.

To obtain the maximum electric field, an appropriate separation distance between the electrodes had to be selected. The separation distances were dictated by the propagation losses introduced by the electrodes near the PE waveguides. The propagation losses with different separation distances between the electrodes at a wavelength of λ = 1.55 μm are shown in Figure 5. The propagation loss increased sharply with the diminishing separation distance. In the following simulation, the separation distance between the electrodes was selected to have a PE waveguide loss of approximately 0.5 dB/cm. For radio frequency (RF) attenuation, thick metals facilitate low-loss RF waveguides. When the electrode thickness was larger than 1 μm, the decrease of the RF attenuation was saturated [12]. To achieve the optimum performance of the electro-optic modulator, the thickness of the electrodes could be selected as 1 μm.

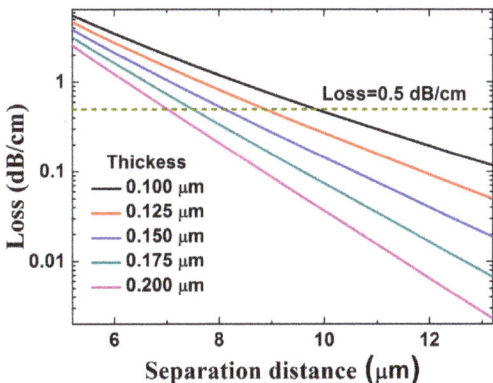

Figure 5. Relationship between the propagation loss and the separation distance.

As shown in Figure 6a,b, the optical field and the electrostatic field were simulated. Figure 6c shows the overlap integral of the optical and electrostatic fields. The PE waveguide had a thickness of 0.15 μm and a width of 1.2 μm, allowing confinement for most of the optical power in the LN core without PE, which was the E-O active material. We could design electrodes to be placed close to the waveguides without substantially bigger optical transmission losses.

To control the optical properties with an external electric signal, the E-O effect or the Pockels effect was used, where the birefringence of the crystal changed proportionally to the applied electric field. A change in the refractive index resulted in a change of the phase of the wave passing through the crystal. If two waves with different phase change were combined, the amplitude modulation could be performed by an interferometer. An important quality factor for the M-ZI modulators was the half-wave voltage (V_π), defined as the required voltage to induce a π-phase difference between the two modulator arms, changing the optical transmission from the maximum to minimum. Figure 7a shows the optical transmission of a device with 1-cm-long microwave strip line electrodes, for which we calculated a low V_π of 10.2 V (the thickness and the width of PE waveguide and the separation distance between electrodes were 0.15, 1.2, and 8.08 μm, respectively). Figure 7b shows that the half-wave voltage length product varied with the thickness of the PE region. The half-wave voltage length product ascended with the increasing thickness of the PE waveguide. Considering the bending loss of the Y branch and the half-wave voltage length product, the most suitable thickness of the PE waveguide was 0.15 μm.

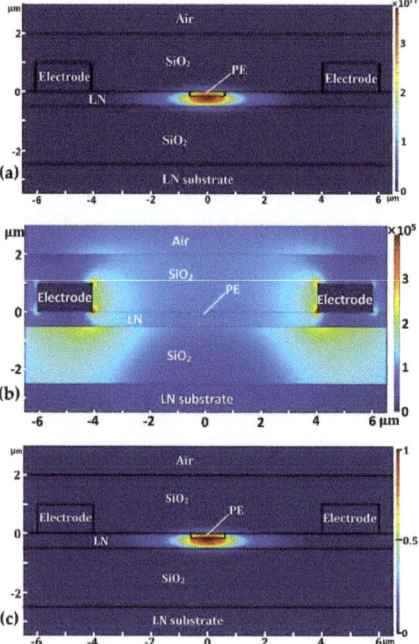

Figure 6. (a) Optical field inside the waveguide. (b) Electrostatic field after a 1 V voltage was applied to the electrodes. (c) Normalized product of the optical and electrical fields.

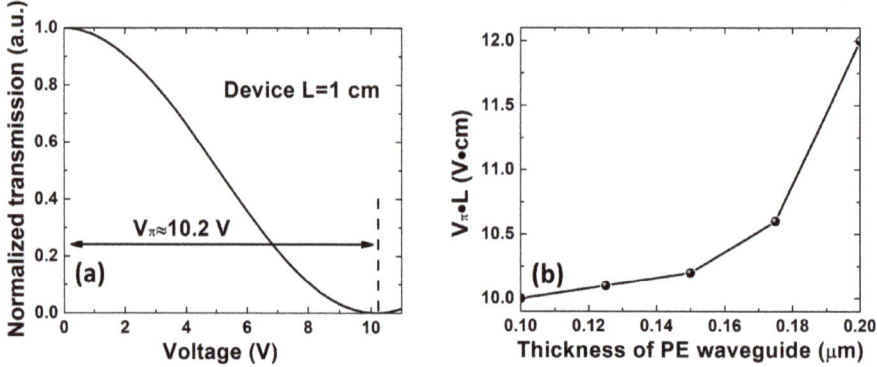

Figure 7. (a) Normalized optical transmission of a 1-cm device as a function of the applied voltage. (b) Half-wave voltage-length product variation with the thickness of the PE waveguide.

The frequency-dependent refractive index mismatch between the optical and RF signals played a key role in the final modulation bandwidth of the modulator [1]. Thanks to the LN thin film structure, the refractive index of the RF and optical modes was well matched in the modulator [12]. This was different to the ordinary bulk LN modulators [35]. By further adjusting the structural parameters of the waveguides and electrodes, the mismatch refractive index between the optical and RF signals should be minimized as much as possible, which should be studied carefully in the future.

4. Conclusions

The full-vectorial finite-difference method was used to calculate the single-mode conditions, mode sizes, and optical power distribution of the PE channel waveguides. The widths of the PE waveguides were optimized to 1.2 µm. The propagation losses of the guided mode at different separation distances between the electrodes were analyzed and discussed. As a very important aspect of the practical application, the half-wave voltages were simulated using FD-BPM. The thickness of the PE waveguides and the separation distances between the electrodes were optimized to 0.15 µm and 8.08 µm, respectively. The optimized value of $V_\pi \cdot L$ was calculated to be 10.2 V·cm.

Author Contributions: B.X. conceived the original idea; H.H. carried out the simulations and wrote the manuscript; H.H., B.X., T.L., G.C. and S.R. contributed the useful and deep discussions, analyzed the data and modified the manuscript.

Funding: This work was supported by the Shenzhen Science and Technology Planning (NO. JCYJ20170818143327496), the Project of Youth Innovative Talents in Higher Education Institutions of Guangdong (NO. 2018KQNCX399), the Foundation of Zibo Vocational Institute (NO. 2018zzzr03), the School City Integration Development Plan of Zibo (NO. 2019ZBXC127), and the Shandong University Science and Technology Planning (NO. J16LN93).

Conflicts of Interest: The authors declare no conflict of interest.

References

1. Honardoost, A.; Safian, R.; Rao, A.; Fathpour, S. High-Speed Modeling of Ultracompact Electrooptic Modulators. *J. Light. Technol.* **2018**, *36*, 5893–5902. [CrossRef]
2. Janner, D.; Tulli, D.; García-Granda, M.; Belmonte, M.; Pruneri, V. Micro-structured integrated electro-optic LiNbO3 modulators. *Laser Photonics Rev.* **2019**, *3*, 301–313. [CrossRef]
3. Chuang, R.W.; Hsu, M.T.; Chang, Y.C.; Lee, Y.J.; Chou, S.H. Integrated multimode interference coupler-based Mach–Zehnder interferometric modulator fabricated on a silicon-on-insulator substrate. *IET Optoelectron.* **2012**, *6*, 147–152. [CrossRef]
4. Chen, H.W.; Kuo, Y.H.; Bowers, J.E. High speed hybrid silicon evanescent Mach-Zehnder modulator and switch. *Opt. Express* **2008**, *16*, 20571–20576. [CrossRef] [PubMed]
5. Ding, J.; Shao, S.; Zhang, L.; Fu, X.; Yang, L. Method to improve the linearity of the silicon Mach-Zehnder optical modulator by doping control. *Opt. Express* **2016**, *24*, 24641–24648. [CrossRef]
6. Kitoh, T. Recent progress on arrayed-waveguide grating multi/demultiplexers based on silica planar lightwave circuits. *Proc. SPIE* **2008**, *7135*, 713503.
7. Maru, K.; Mizumoto, M.; Uetsuka, H. Demonstration of Flat-Passband Multi/Demultiplexer Using Multi-Input Arrayed Waveguide Grating Combined with Cascaded Mach-Zehnder Interferometers. *J. Light. Technol.* **2007**, *25*, 2187–2197. [CrossRef]
8. Cahill, L.W. The modelling of integrated optical power splitters and switches based on generalised Mach-Zehnder devices. *Opt. Quantum Electron.* **2004**, *36*, 165–173. [CrossRef]
9. Cherchi, M. Design scheme for Mach–Zehnder interferometric coarse wavelength division multiplexing splitters and combiners. *J. Opt. Soc. Am. B* **2006**, *23*, 1752–1756. [CrossRef]
10. Weis, R.S.; Gaylord, T.K. Lithium niobate: Summary of physical properties and crystal structure. *Appl. Phys. A* **1985**, *37*, 191–203. [CrossRef]
11. Wang, C.; Zhang, M.; Chen, X.; Bertrand, M.; Shams-Ansari, A.; Chandrasekhar, S.; Winzer, P.; Lončar, M. Integrated lithium niobate electro-optic modulators operating at CMOS-compatible voltages. *Nature* **2018**, *562*, 101–104. [CrossRef] [PubMed]
12. He, M.; Xu, M.; Ren, Y.; Jian, J.; Ruan, Z.; Xu, Y.; Gao, S.; Sun, S.; Wen, X.; Zhou, L.; et al. High-performance hybrid silicon and lithium niobate Mach–Zehnder modulators for 100 Gbit s^{-1} and beyond. *Nat. Photonics* **2019**, *13*, 359–364. [CrossRef]
13. Weigel, P.O.; Zhao, J.; Fang, K.; Al-Rubaye, H.; Trotter, D.; Hood, D.; Mudrick, J.; Dallo, C.; Pomerene, A.T.; Starbuck, A.L.; et al. Bonded thin film lithium niobate modulator on a silicon photonics platform exceeding 100 GHz 3-dB electrical modulation bandwidth. *Opt. Express* **2018**, *26*, 23728–23739. [CrossRef] [PubMed]
14. Mercante, A.J.; Shi, S.; Yao, P.; Xie, L.; Weikle, R.M.; Prather, D.W. Thin film lithium niobate electro-optic modulator with terahertz operating bandwidth. *Opt. Express* **2018**, *26*, 14810–14816. [CrossRef]

15. Rao, A.; Patil, A.; Rabiei, P.; Honardoost, A.; Desalvo, R.; Paolella, A.; Fathpour, S. High-performance and linear thin-film lithium niobate Mach–Zehnder modulators on silicon up to 50 GHz. *Opt. Lett.* **2016**, *41*, 5700–5703. [CrossRef]
16. Poberaj, G.; Hu, H.; Sohler, W.; Günter, P. Lithium niobate on insulator (LNOI) for micro-photonic devices. *Laser Photonics Rev.* **2012**, *6*, 488–503. [CrossRef]
17. Han, H.; Cai, L.; Hu, H. Optical and structural properties of single-crystal lithium niobate thin film. *Opt. Mater.* **2015**, *42*, 47–51. [CrossRef]
18. Xiang, B.; Guan, J.; Jiao, Y.; Wang, L. Fabrication of ion-sliced lithium niobate slabs using helium ion implantation and Cu–Sn bonding. *Phys. Status Solidi (A)* **2014**, *211*, 2416–2420. [CrossRef]
19. Hu, H.; Ricken, R.; Sohler, W. Lithium niobate photonic wires. *Opt. Express* **2009**, *17*, 24261–24268. [CrossRef]
20. Zhang, M.; Wang, C.; Cheng, R.; Shams-Ansari, A.; Lončar, M. Monolithic ultra-high-Q lithium niobate microring resonator. *Optica* **2017**, *4*, 1536–1537. [CrossRef]
21. Honardoost, A.; Juneghani, F.A.; Safian, R.; Fathpour, S. Towards subterahertz bandwidth ultracompact lithium niobate electrooptic modulators. *Opt. Express* **2019**, *27*, 6495–6501. [CrossRef] [PubMed]
22. Cai, L.; Mahmoud, A.; Piazza, G. Low-loss waveguides on Y-cut thin film lithium niobate: Towards acousto-optic applications. *Opt. Express* **2019**, *27*, 9794–9802. [CrossRef] [PubMed]
23. Cai, L.; Kong, R.; Wang, Y.; Hu, H. Channel waveguides and y-junctions in x-cut single-crystal lithium niobate thin film. *Opt. Express* **2015**, *23*, 29211–29221. [CrossRef] [PubMed]
24. Wu, R.; Wang, M.; Xu, J.; Qi, J.; Chu, W.; Fang, Z.; Zhang, J.; Zhou, J.; Qiao, L.; Chai, Z.; et al. Long Low-Loss-Litium Niobate on Insulator Waveguides with Sub-Nanometer Surface Roughness. *Nanomaterials* **2018**, *8*, 910. [CrossRef] [PubMed]
25. Ganshin, V.A.; Korkishko, Y.N. Proton exchange in lithium niobate and lithium tantalate single crystals: Regularities and specific features. *Phys. Status Solidi (A)* **1990**, *119*, 11–25. [CrossRef]
26. Canali, C.; Carnera, A.; Mea, G.D.; Mazzoldi, P.; Shukri, S.M.A.; Nutt, A.C.G.; Rue, R.M.D.L. Structural characterization of proton exchanged LiNbO$_3$ optical waveguides. *J. Appl. Phys.* **1986**, *59*, 2643–2649. [CrossRef]
27. Xu, C.L.; Huang, W.P.; Stern, M.S.; Chaudhuri, S.K. Full-vectorial mode calculations by finite difference method. *IEE Proc. Optoelectron.* **1994**, *141*, 281–286. [CrossRef]
28. Scarmozzino, R.; Gopinath, A.; Pregla, R.; Helfert, S. Numerical techniques for modeling guided-wave photonic devices. *IEEE J. Sel. Top. Quantum Electron.* **2000**, *6*, 150–162. [CrossRef]
29. Scarmozzino, R.; Osgood, R.M. Comparison of finite-difference and Fourier-transform solutions of the parabolic wave equation with emphasis on integrated-optics applications. *J. Opt. Soc. Am. A* **1991**, *8*, 724–731. [CrossRef]
30. Han, H.; Xiang, B.; Zhang, J. Simulation and analysis of single-mode microring resonators in lithium niobate thin film. *Crystals* **2018**, *8*, 342. [CrossRef]
31. Méndez, A.; Paliza, G.D.L.; Garcia-Cabanes, A.; Cabrera, J.M. Comparison of the electro-optic coefficient r33 in well-defined phases of proton exchanged LiNbO3 waveguides. *Appl. Phys. B* **2001**, *73*, 485–488. [CrossRef]
32. Schlarb, U.; Betzler, K. A generalized sellmeier equation for the refractive indices of lithium niobate. *Ferroelectrics* **1994**, *156*, 99–104. [CrossRef]
33. Lumerical Solutions. Available online: http://www.lumerical.com/ (accessed on 25 August 2019).
34. Chung, Y.; Dagli, N. An assessment of finite difference beam propagation method. *IEEE J. Quantum Electron.* **1990**, *26*, 1335–1339. [CrossRef]
35. Wooten, E.; Kissa, K.; Yi-Yan, A.; Murphy, E.; Lafaw, D.; Hallemeier, P.; Maack, D.; Attanasio, D.; Fritz, D.; McBrien, G.; et al. A review of lithium niobate modulators for fiber-optic communications systems. *IEEE J. Quantum Electron.* **2000**, *6*, 69–82. [CrossRef]

 © 2019 by the authors. Licensee MDPI, Basel, Switzerland. This article is an open access article distributed under the terms and conditions of the Creative Commons Attribution (CC BY) license (http://creativecommons.org/licenses/by/4.0/).

Article

An Integrated Photonic Electric-Field Sensor Utilizing a 1 × 2 YBB Mach-Zehnder Interferometric Modulator with a Titanium-Diffused Lithium Niobate Waveguide and a Dipole Patch Antenna

Hongsik Jung

Department of Electronic and Electrical Fusion Engineering, Hongik University, Sejong 30016, Korea; hsjung@hongik.ac.kr; Tel.: +82-44-860-2532

Received: 29 July 2019; Accepted: 31 August 2019; Published: 2 September 2019

Abstract: We studied photonic electric-field sensors using a 1 × 2 YBB-MZI modulator composed of two complementary outputs and a 3 dB directional coupler based on the electro-optic effect and titanium diffused lithium–niobate optical waveguides. The measured DC switching voltage and extinction ratio at the wavelength 1.3 μm were ~16.6 V and ~14.7 dB, respectively. The minimum detectable fields were ~1.12 V/m and ~3.3 V/m, corresponding to the ~22 dB and ~18 dB dynamic ranges of ~10 MHz and 50 MHz, respectively, for an rf power of 20 dBm. The sensor shows an almost linear response to the applied electric-field strength within the range of 0.29 V/m to 29.8 V/m.

Keywords: photonic electric-field sensor; titanium diffused optical channel waveguide; lithium–niobate electro-optic effect; Y-fed balanced-bridge Mach-Zehnder interferometer (YBB-MZI)

1. Introduction

Electric-field sensors that exhibit wide, flat frequency response characteristics are important tools for electromagnetic compatibility and interference (EMC/EMI) measurements, high-frequency electronic circuit analysis, medical equipment field observation, radio-frequency reception, and high-power microwave detection. The importance of these sensors is increasing as mobile multimedia communications develops [1,2]. It is necessary to accurately evaluate the strength and distribution of electromagnetic fields surrounding electronic equipment to estimate electromagnetic compatibility. The requirements for electric-field sensors based on the important applications mentioned above are as follows: their wide frequency bandwidth and large dynamic range; their high spatial resolution and low interference to the original field; and their high stability and accuracy.

Even though a variety of sensing modules have been developed, those with photonic links reduce or eliminate some of the inaccuracies and systematic errors that affect measurement techniques using conventional EM-field sensors. They provide electrical isolation, which eliminates ground loops and common-mode electrical pickup between the sensor head and the electronics module. Moreover, the optical fibers and dielectric components produce minimal field distortion. In addition, they can preserve both the phase and amplitude of high-frequency fields with good fidelity and low losses. The development of optical-fiber and optoelectronic components for the telecommunications industry has made it possible to implement photonic sensors that are accurate and convenient to use.

In particular, titanium-diffused lithium niobate (Ti: $LiNbO_3$) waveguide devices are suitable for electric-field detection since their sensors will not perturb the field to be measured. A linear modulator that is passively biased to the optimal linear operating point is required. This has been demonstrated for asymmetric Mach–Zehnder interferometers (MZIs) and 1 × 2 directional couplers. The former devices have an intrinsic bias of π/2, where a geometrical path length difference of a quarter of a wavelength is required between the two arms [3–7]. However, it is not easy to obtain optimal operation through path length difference alone, because of fabrication tolerances.

Moreover, the latter device is automatically biased to the optimal 3 dB operating point due to its symmetrical structure, and it provides a greater tolerance in the fabrication process than an asymmetric Mach–Zehnder interferometric optical modulator [8–11]. However, the complexity of the transfer function makes it impossible to utilize the sensor in a specific range (namely, the ratio of interaction length to conversion length).

In contrast, a Y-fed balanced-bridge MZI modulator (YBB-MZI) consists of a 3 dB coupler at the output, with two complementary output waveguides [12–19]. This type of modulator provides a well-defined transfer function for the output optical power versus the detected electric-field intensity and can be automatically biased at the optimum 3 dB operating point due to its symmetrical structure, which offers a more tolerant design in the fabrication process than Mach–Zehnder interferometric modulators or 1 × 2 directional couplers, which are asymmetrical. A mono-shield gold electrode structure was applied in YBB-MZI by the Tsinghua University group to detect a very high electric field [12]. Moreover, the minimum electric-field strength that can be sensed is determined by the relative intensity noise (RIN) of the laser diode. Therefore, a YBB-MZI configuration was proposed by a German group to reduce RIN and improve sensitivity with a balanced optical receiver [16].

In this paper, we provide the quantitative theory of a YBB-MZI modulator and report on a fabricated Ti: LiNbO$_3$ YBB-MZI modulator operating at a 1.3 μm wavelength. We also present, in detail, the fabrication process and design parameters, as well as the optical and electrical performances of a packaged electric-field sensor with a diploe patch antenna, including the minimum detectable electric-field intensity, dynamic range, and sensitivity.

2. Theory, Fabrication, and Performance of a Ti: LiNbO$_3$ YBB-MZI Modulator

2.1. Device Theory

The YBB-MZI modulator consists of a 3 dB directional coupler at the output and has two complementary output waveguides, as shown in Figure 1. A dipole patch antenna was placed around the arm of the MZI structure to detect the electric field.

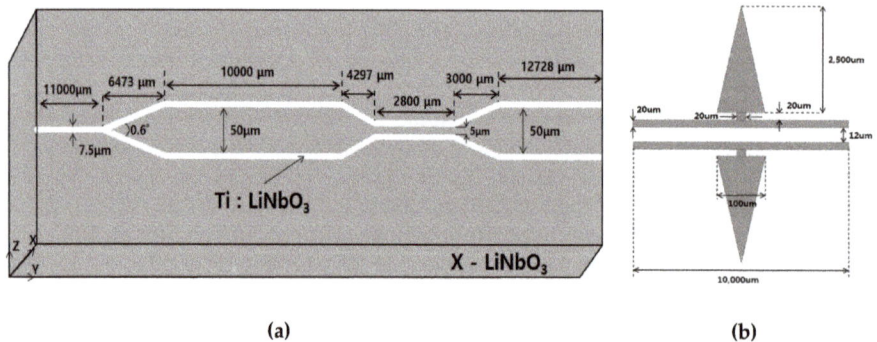

Figure 1. Schematic diagrams and dimensions of (**a**) a Ti: LiNbO$_3$ 1 × 2 Y-fed balanced-bridge Mach–Zehnder interferometer (YBB-MZI) modulator and (**b**) a dipole patch antenna.

The operating characteristics of a 2 × 2 directional coupler are represented by the coupling length L_c, the coupling coefficient κ, and the wavenumber β of the waveguide. If the transmission loss is ignored, the transfer matrix of a directional coupler is expressed by [20]:

$$\begin{pmatrix} \overline{E_{01}} \\ \overline{E_{02}} \end{pmatrix} = e^{-j\beta L_c} \begin{pmatrix} \cos \kappa L_c & -j \sin \kappa L_c \\ -j \sin \kappa L_c & \cos \kappa L_c \end{pmatrix} \begin{pmatrix} \overline{E_{i1}} \\ \overline{E_{i2}} \end{pmatrix}. \tag{1}$$

where E_{i1}, E_{i2}, and E_{01}, E_{02} are the input and output optical modes, respectively. The incident single-mode optical-wave is equally divided in two by a 3 dB power splitter located at the input stage and can be expressed as follows:

$$\overline{E_{i1}}, \overline{E_{i2}} = \frac{1}{\sqrt{2}}e^{-j\theta} \tag{2}$$

where θ is the initial phase.

The dipole patch antenna with an electrode, as shown in Figure 1b, creates an electric field on one of the two arms of the MZI, which eventually induces a change of the refractive index and an unbalanced modulation. Before going into the output directional coupler, the optical wave in the two arms has an extrinsic phase mismatch $\Phi(E_e)$ due to the detected electric field. This phase mismatch $\Phi(E_e)$ can be expressed as

$$\Phi(E_e) = \pm\frac{\pi}{\lambda}n_e^3 \gamma_{33} \Gamma l_e E_e \tag{3}$$

where l_e is the length of the electrode connected to the dipole patch antenna, r_{33} is the electro-optic coefficient of lithium niobate (~30 pm/V), λ is the optical wavelength, n_e is the extraordinary refractive index of lithium niobate, E_e is the electric-field strength through the waveguide, and Γ ($0 < \Gamma < 1$) is the overlap integral between the applied electrical field and the optical wave. Therefore, the optical-wave going into the output coupler can be represented as

$$\overline{E}_{i1} = \frac{1}{\sqrt{2}}e^{-j(\theta+\Phi(E_e))} \tag{4a}$$

$$\overline{E}_{i2} = \frac{1}{\sqrt{2}}e^{-j\theta}. \tag{4b}$$

Combining (1) with (4), the output power of the YBB-MZI modulator is expressed as

$$P_{01} = \frac{1}{2}[1 + sin(\pi y) \cdot sin(\pi x)] = \frac{1}{2}[1 + sin(2kL_c) \cdot sin(\Phi(E_e))] \tag{5a}$$

$$P_{02} = \frac{1}{2}[1 - sin(\pi y) \cdot sin(\pi x)] = \frac{1}{2}[1 - sin(2kL_c) \cdot sin(\Phi(E_e))] \tag{5b}$$

where $x = \Phi(E_e)/\pi$ is the normalized phase-mismatch, $y = L_c/l_c$ is the normalized coupling length, and $l_c = \pi/2k$ is the coupling conversion length.

The output intensity P_{o1} is simulated and plotted for the YBB-MZI electric-field sensor, as shown in Figure 2. The YBB-MZI sensor shows a sinusoidal transfer function for different y-values. The value of y only affects the extinction ratio, which can be represented as $sin(\pi y)$. For most cases (where $sin(\pi y) \neq 0$), the transfer function is acceptable as the extinction ratio only impacts the E-field measurement sensitivity. To support the maximum sensitivity, the coupling length should satisfy the condition

$$sin(2k \cdot L_c) = 1, \tag{6}$$

where $k \cdot L_c = \frac{(2n+1)\pi}{4}$, ($n = 0, 1, 2, \ldots$).

2.2. Designs and Fabrication

Using single-mode Ti:LiNbO$_3$ channel waveguides, a symmetric 1 × 2 YBB-MZI modulator with a dipole patch antenna was designed for operation at a wavelength of ~1.3 µm in an x-cut, y-propagating LiNbO$_3$ substrate, as shown in Figure 1. The device consists of a Y-branch splitter, a phase modulator, and a directional coupler. The entire device's structure is similar to that of a Mach–Zehnder interferometer with two output ports. The waveguide width is 7.5 µm for single-mode operation, and the splitting angle of the Y-branch is 0.6° for decreasing the propagation loss as low as possible and for fabrication tolerance. The gap interval between the two adjacent waveguides of the directional coupler and the parallel coupling length are 5 µm and 2.8 mm, respectively, to split the optical power equally into two output channels with a nominal coupling constant-length product, κ·Lc of π/4. The interval between the inner edges of the two output waveguides is 50 μm, thereby preventing optical power coupling between the two output channels. As shown in Figure 1b, the gap and length of the modulation electrode connected to the dipole patch antenna are 12 μm and 10 mm, respectively. The results of the BPM-CAD 3D simulation of the optical wave propagating through the YBB-MZI modulator are shown in Figure 3 [21]. When no voltage was applied, the two intensity profiles were approximately identical, with ~1% or lower accuracy because of the nearly equal intensity splitting, as shown in Figure 3a. Therefore, the YBB-MZI modulator was intrinsically set at the 3 dB half-power point. While the driving voltage increased to 5 V and 10 V, the light in the lower branch of the device was coupled with the upper branch, where the light intensity of the lower branch decreased and the intensity of the upper branch increased, as shown in Figure 3b,c. When 10 V was applied, the light of the lower branch almost disappeared, and the light intensity of the upper branch reached the highest level. Therefore, it could be theoretically confirmed that the switching voltage required to modulate the light intensity of either branch from a bar state (maximum intensity) to a cross state (minimum intensity) was ~10 V.

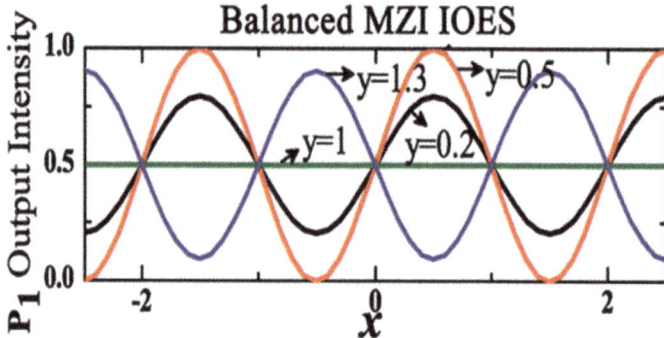

Figure 2. Simulation results for the light output intensity versus driving voltage with y = 0.2, 0.5, 1, and 1.3.

An investigation of the formation of optical waveguides in $LiNbO_3$ by metal ion diffusion indicated an increase or decrease in the refractive index depending on the valence of the in-diffused ion. Higher-valence ions such as Ti^{3+}, Fe^{3+}, and Cr^{3+} increase both the ordinary and extraordinary indices. It appears that lower-valence ions replace Li^+ sites, while higher-valence ions replace Nb^{5+} sites. Experimental results indicated that the in-diffused Ti metal in $LiNbO_3$ was all tetravalent (i.e., Ti atoms are fully ionized). There are no electrons in particularly filled d-orbitals to absorb the electromagnetic energy at visible wavelengths. This explains the measurement of low losses of waveguides fabricated by Ti diffusion into $LiNbO_3$ [22–24]. The dominant sources of waveguide loss are scattering from $LiNbO_3$ surface imperfections due to diffusion and possibly absorption by the metal ions.

The 1 × 2 YBB-MZI waveguide structure, as shown in Figure 1a, was fabricated on an x-cut, 3-inch, 1-mm-thick $LiNbO_3$ wafer as the substrate using UV photolithography and thermal diffusion. First, a 1050-Å-thick Ti-film on the $LiNbO_3$ substrate was deposited by an e-beam evaporator, and then the desired Ti-film patterns with 7.5 μm widths were formed by photolithography and the wet-etching process, followed by thermal diffusion for 8 hours at 1050 °C in wet-ambient. The resulting Ti-diffused channel waveguides grew to a thickness two or three times that of the Ti-film stripe. Such surface growth makes it easy to observe Ti-diffused waveguides with a microscope, as shown in Figure 4. Furthermore, the diffused waveguide has Gaussian index profiles in its depths. The effective index increases in linear proportion to the Ti film's thickness. This feature indicates that the propagation constant of the fundamental mode can easily be controlled by changing the film thickness alone.

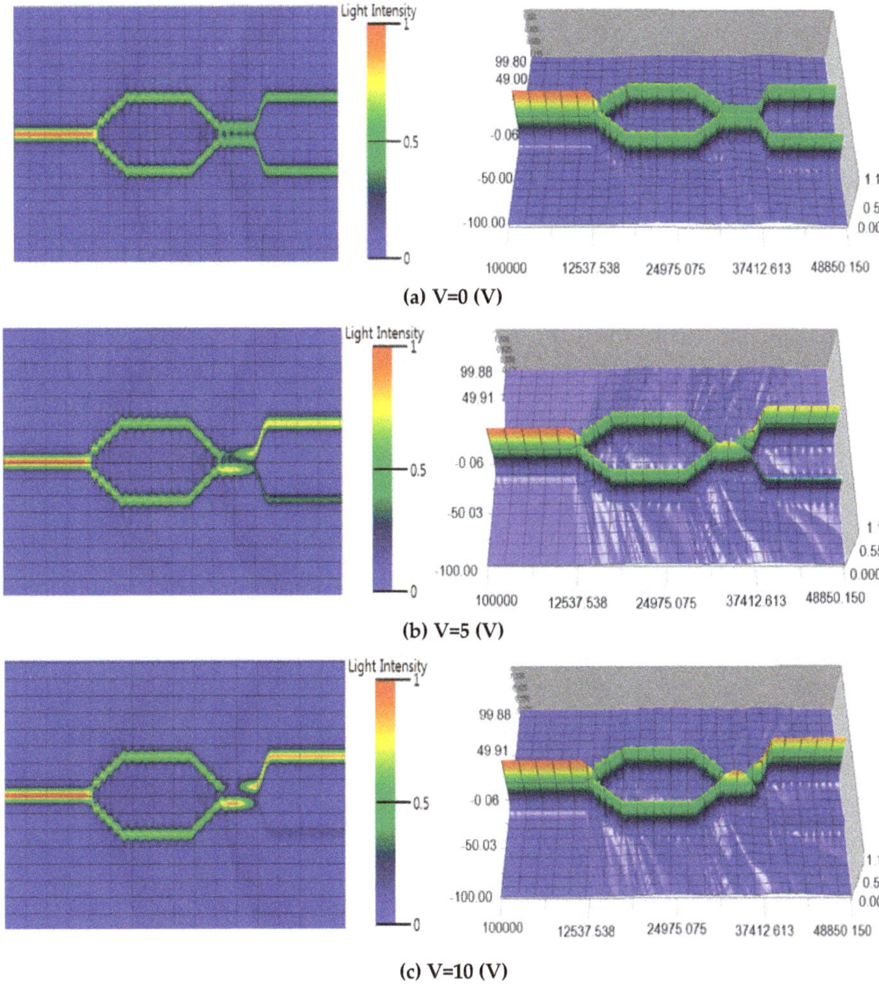

Figure 3. Three-dimensional BPM-CAD simulation results with the following applied voltages: (**a**) 0 V, (**b**) 5 V, and (**c**) 10 V.

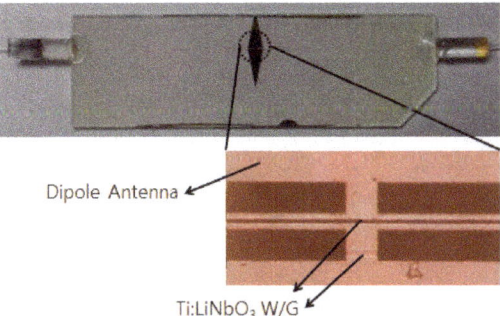

Figure 4. Photograph of the implemented device with pig-tailed optical fibers. W/G is the abbreviation for waveguide.

The waveguide edges were optically polished to allow butt-coupling and pig-tailing. A silicon dioxide buffer layer with a thickness of ~3000 Å was deposited on the substrate using an electron beam and 99.99% pure SiO_2 pellets to reduce the propagation loss due to the absorption of the light wave of the antenna's metal. An aluminum dipole patch antenna and electrode ~5000 Å thick (as shown in Figure 1b) were formed along one of the two arms of the YBB-MZI to allow sensing of the electric field. A polarization-maintaining single-mode optical fiber and multi-mode fiber were attached to the input and output waveguides, respectively. Figure 4 shows a photograph of the implemented device with the attached optical fibers and a dipole patch antenna. The insertion loss of the device, including the input/output fiber, was measured to be about 11.7 dB, which includes the fiber-connector loss, pig-tailing loss, mode-mismatch loss, and propagation loss of the waveguides.

2.3. Performance Evaluations

The fabricated device without an attached optical fiber was first tested by applying DC voltages. The device's performance and characterization were observed using a tunable laser with butt-coupling at a wavelength of 1.3. The TE-polarized input light was butt-coupled to the devices, collected at the output by a microscope's objective lens, and focused onto a photo-detector for measurement. TE or TM polarized light was selected by properly adjusting a fiber optic polarization controller. We first observed the single-mode propagation for TE polarization in the 1 × 2 YBB-MZI modulator.

It was observed that when the voltage was not applied, the two outputs of the device were almost the same. The voltage required to switch either output light power from a bar state (maximum intensity) to a cross state (minimum intensity) was measured to be ~-16.6 V, which corresponds to a ~14.7 dB extinction ratio. Figure 5 shows the optical output power versus the DC voltage measured by an optical power meter and shows a slightly asymmetric DC output characteristic, as well as a switching voltage of ~16.6 V, as mentioned previously. The AC modulation responses of the two outputs versus the driving sinusoidal voltage were further measured in Figure 6, where the optical signals are below the sinusoidal curve and the applied ac voltage signal is above the curve, at a frequency of ~1 kHz (5 V/div). The power of the two outputs was confirmed to be nearly equal, and the periodic exchange of output power in the two outputs expresses a good inverse relationship in the output sinusoidal curves. The slightly skewed and flattened shape in the optical response in Figure 6 was observed due to the imperfect single mode waveguide and the out-diffusion that occurred in the diffusion process.

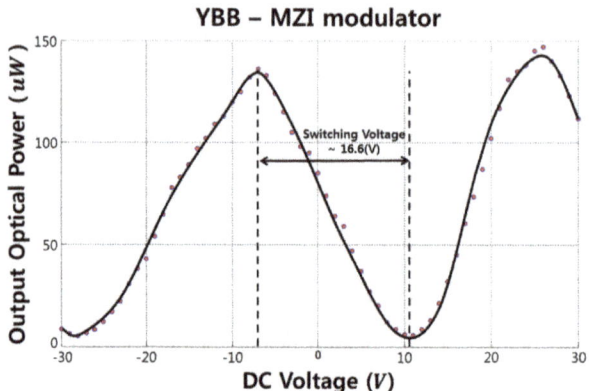

Figure 5. Measured optical output power intensity versus the applied DC voltage.

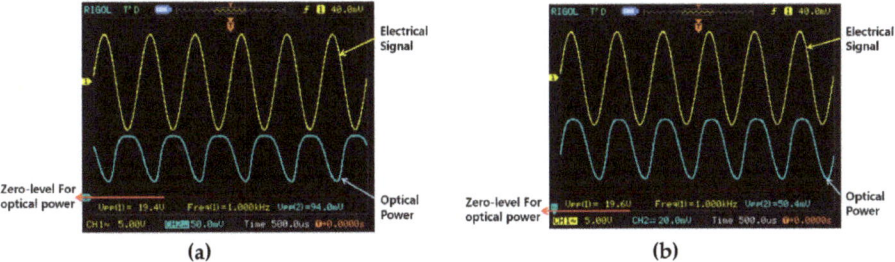

Figure 6. The 1 kHz ac modulation responses at the (**a**) upper and (**b**) lower output port, as shown in Figure 1a.

3. Measurement and Experimental Results

3.1. Experimental Setup

To measure the frequency responses and the minimum measurable field strength of the device, frequency tests were performed utilizing a tunable laser at a wavelength of 1.3 μm. The input optical power was about 1.4 mW. A detailed diagram of the measurement setup is shown in Figure 7. The device was tested in a uniform electric-field environment by placing it in a Transverse Electro Magnetic (TEM) cell (Tescom TC-5010A), as shown in Figure 7, where the TEM cell is utilized to generate accurate electro-magnetic (EM) waves over a wide frequency range. EM waves generated in the cell are transmitted in the transverse mode and have similar characteristics to the plane-wave. The optical fibers penetrating through the slanted wall of the TEM cell and were connected to the laser and photodetector using an FC/PC fiber-optic connector. The applied electric-field strength was calculated using the output level and the size of the TEM cell. The −20~+20 dBm (10 μW~100 mW) rf input power to the TEM cell corresponds to the electric-field strength from 0.293 V/m to 23.2 V/m. Due to the high permittivity ($\varepsilon \approx 35$) of $LiNbO_3$, the substantial electric-field intensity experienced on the sensor substrate (23.2 V/m) corresponds to 0.66 V/m in the TEM cell. The rf power propagates through the TEM cell in the same direction as the light passing through the optical fibers and the sensor [9].

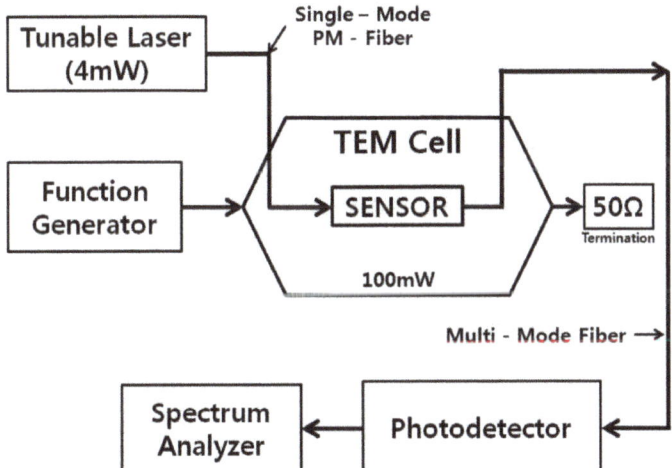

Figure 7. Block diagram of the measurement setup for electric-field sensing and the evaluation of the frequency response. TEM and PM are abbreviations for Transverse Electro Magnetic and polarization maintaining, respectively.

3.2. Test Results and Discussions

Figure 8 shows the spectrum-analyzer outputs for an input of 20 dBm to the TEM cell at frequencies of 10, 50, and 70 MHz, respectively. The rf power detected at the photodetector was measured to be −101.5, −110.9, and −122.2 dBm, as shown in Figure 8, and the noise floor was measured to be about −130 dBm at the same frequencies. The internal electric field of 29.8 V/m in the TEM cell produced an SNR of 28.5, 19.1, and 7.8 dB at each frequency. Therefore, the minimum detectable electric-fields are ~1.12, ~3.3, and ~12.13 V/m, respectively, at those three frequencies, based on the equation $E_{min} = 29.8 \times 10^{(-SNR/20)}$ V/m.

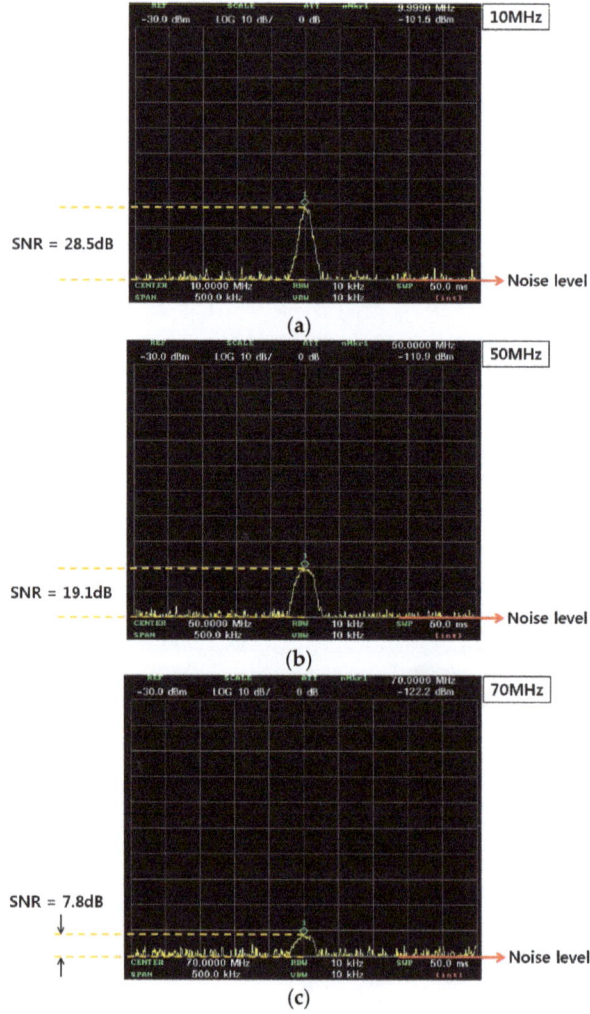

Figure 8. The detected rf spectra of (**a**) 10 MHz, (**b**) 50 MHz, and (**c**) 70 MHz rf input signals into the TEM cell, with a power level of 100 mW.

Figure 9 shows the sensitivity curves at rf frequencies of 10 MHz, 50 MHz, and 100 MHz. We can confirm that the graph shows almost linear response characteristics from the applied electric-field intensity from 0.293 V/m to 23.2 V/m. Even though some data are off the linear response line, they remain very close. The device also shows a dynamic range of about ~22, ~18, and ~12 dB at frequencies of 10, 50, and 100 MHz, respectively. Figure 10 shows the photodetector power at different electric-field intensities.

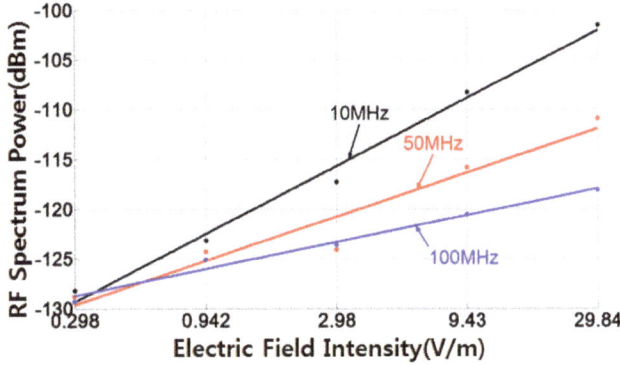

Figure 9. The photo-detected signal power versus the electric-field strength in the TEM cell at different frequencies.

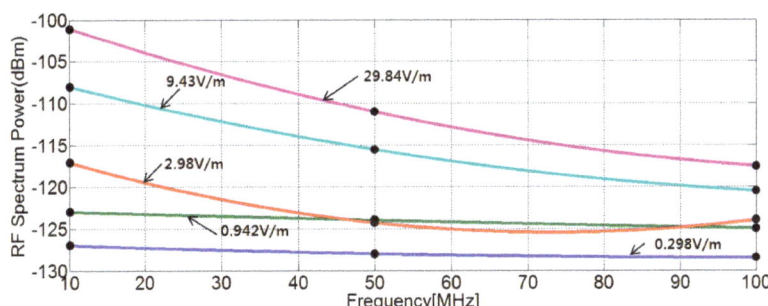

Figure 10. The photo-detected signal power versus frequency at different electric-field strengths in the TEM cell.

Figure 11 illustrates the frequency response of the sensor measured with 20 dBm of rf input power applied to the TEM cell. This figure shows a nearly flat frequency response from 1 MHz to ~50 MHz. The cut-off high frequency of the device is derived from the series-coupled time constant of the electrode resistance and the structural and packaged capacitances of the device. Therefore, a much higher cutoff frequency can be expected when a metal material with a higher coefficient of conductivity, such as gold instead of aluminum, is applied to an electrode.

So far, the theoretical analysis and experimental results have confirmed that an electric-field sensor based on YBB-MZI exhibits a superior 3 dB optical bias and simple sinusoidal transfer characteristics. Regardless of the refractive index of the optical waveguide, a 3 dB optical bias was obtained because of the perfect symmetry of the two arms that make up the YBB-MZI. However, in the case of a conventional MZI, a 3 dB optical bias can be realized by the optical path difference between the two arms. Moreover, the optical bias depends on both the optical path difference and the effective refractive index of the waveguide, which is especially affected by fabrication parameters, such as titanium thickness, diffusion time, temperature, and ambience. Therefore, the YBB-MZI structure allows much better control of the optical bias than does a conventional MZI.

Figure 11. The frequency response of the sensor.

4. Conclusions

We have demonstrated a photonic electric-field sensor utilizing a 1 × 2 electro-optic Ti: LiNbO$_3$ Y-fed balanced bridge Mach–Zehnder Interferometric modulator, which provides the unique characteristic of an intrinsic 3 dB operating point, due to its symmetrical geometry. The theoretical analysis demonstrates that the YBB-MZI structure inherits advantages from both conventional MZI and directional coupler structures: namely, a sinusoidal transfer function and a better optical bias control. The sensors were designed and fabricated with a 49 × 15 × 1 mm size and operated at a wavelength of 1.3 µm. We observed a dc switching voltage of ~16.6 V and an extinction ratio of ~14.7 dB. The minimum detectable electric-field strengths for this device were ~1.12 V/m and ~3.3 V/m, corresponding to a dynamic range of about ~22 dB and ~18 dB at frequencies of 10 MHz and 50 MHz, respectively. The sensor exhibits a nearly linear response to an applied electric-field intensity from 0.29 V/m to 29.8 V/m.

In the future, further work on electric-field sensors will be needed to improve sensitivity, operational stability, response speed, detectable frequency range, and encapsulation. To realize a high sensitivity, it is necessary to suppress the noise in the laser diode and the photodetector as much as possible while improving the performance efficiency of the YBB-MZI modulator. The sensitivity limited by shot-noise can be improved by suppressing the relative intensity noise (RIN) of the laser diode as much as possible in the photodetector, and it is possible to configure the balanced detection receiver by combining a YBB-MZI modulator and a balanced photodetector. Since the sensitivity of electric-field sensors utilizing various Ti: LiNbO$_3$-integrated optical modulators is greatly affected by the structures of electrodes and antennas, the performance of sensors based on various electrode structures and antennas (such as dipole antennas, loop antennas, and segmented patch antennas) should be compared and discussed together.

Funding: This research was supported by the Basic Science Research Program through the national Research Foundation of Korea (NRF: 2018049908) and funded by the Ministry of Education, Science and Technology.

Conflicts of Interest: The authors declare no conflict of interest.

References

1. Kuwabara, N.; Tajima, K.; Kobayashi, R.; Amemiya, F. Development and analysis of electric field sensor using LiNbO$_3$ optical modulator. *IEEE Trans. Electromagn. Compat.* **1992**, *34*, 391–396. [CrossRef]
2. Zeng, R.; Wang, B.; Niu, B.; Yu, Z. Development and Application of Integrated Optical Sensors for Intense E-field Measurement. *Sensors* **2012**, *12*, 11406–11434. [CrossRef] [PubMed]
3. Jung, H.S. Photonic Electric-Field Sensor Utilizing an Asymmetric Ti:LiNbO$_3$ Mach-Zehnder Interferometer with a Dipole Antenna. *Fiber Integr. Opt.* **2012**, *31*, 343–354. [CrossRef]

4. Lee, T.H.; Hwang, F.T.; Shay, W.T.; Lee, C.T. Electromagnetc Field Sensor Using Mach-Zehnder Waveguide Modulator. *Microw. Opt. Technol. Lett.* **2006**, *48*, 1897–1899. [CrossRef]
5. Naghski, D.H.; Boyd, J.T.; Jackso, H.E.; Sriram, S.; Kingsley, S.A.; Latess, J. An Integrated Photonic Mach-Zehnder Interferometer with No Electrodes for Sensing Electric Fields. *J. Lightwave Technol.* **1994**, *12*, 1092–1098.
6. Meier, T.; Kostrzewa, C.; Petermann, K.; Schuppert, B. Integrated optical E-field probes with segmented modulator electrodes. *J. Lightwave Technol.* **1994**, *12*, 1497–1503. [CrossRef]
7. Bulmer, C.H.; Burns, W.K. Linear interferometric modulators in Ti:LiNbO$_3$. *J. Lightwave Technol.* **1984**, *2*, 512–521. [CrossRef]
8. An, D.; Shi, Z.; Sun, L.; Taboada, J.M.; Zhou, Q.; Lu, X. Polymeric electro-optic modulator based on 1×2 Y-fed directional coupler. *Appl. Phys. Lett.* **2005**, *76*, 98–104. [CrossRef]
9. Jung, H.S. Electro-optic electric-field sensors utilizing Ti:LiNbO$_3$ 1×2 directional coupler with dipole antennas. *Opt. Eng.* **2013**, *52*, 064402. [CrossRef]
10. Thackara, J.I.; Chon, J.C.; Bjorklund, G.C.; Volksen, W.; Burland, D.M. Polymeric electro-optic Mach–Zehnder switches. *Appl. Phys. Lett.* **1995**, *67*, 3874–3876. [CrossRef]
11. Howerton, M.M.; Bulmer, C.H.; Burns, W.K. Linear 1×2 directional coupler for electromagnetic field detection. *Appl. Phys. Lett.* **1988**, *52*, 1850–1852. [CrossRef]
12. Zeng, R.; Wang, B.; Yu, Z.; Ben Niu, B.; Hua, Y. Integrated optical E-field sensor based on balanced Mach-Zehnder inferometer. *Opt. Eng.* **2011**, *50*, 114404. [CrossRef]
13. Twu, R.C. Zn-Diffused 1×2 Balanced-Bridge Optical Switch in a Y-cut Lithium Niobate. *IEEE Photonics Tech. Lett.* **2007**, *19*, 1269–1271.
14. Chiba, A.; Kawanish, T.; Sakamoto, T.; Higuma, K.; Takada, K.; Izutsu, M. Low-Crosstalk Balanced Bridge Interferometric-Type Optical Switch for Optical Signal Routing. *IEEE J. Sel. Top. Quant.* **2013**, *19*, 3400307. [CrossRef]
15. Lee, M.H.; Min, Y.H.; Ju, J.J.; Do, Y.; Park, S.K. Polymeric electrooptic 2 × 2 switch consisting of birfurcation optical active waveguides and a Mach-Zehnder interferometer. *IEEE J. Sel. Top. Quant* **2013**, *7*, 812–818.
16. Schwerdt, M.; Berger, J.; Schuppert, B.; Petermann, K. Integrated Optical E-Field Sensors with a Balanced Detection Scheme. *IEEE Trans. Electromagn. Compat.* **1997**, *39*, 386–390. [CrossRef]
17. Ramaswamy, V.; Divino, M.D.; Standley, R.D. Balanced bridge modulator switch using Ti-diffused LiNbO$_3$ strip waveguides. *Appl. Phys. Lett.* **1978**, *32*, 644–646.
18. Liu, P.L.; Li, B.J.; Trisno, Y.S. In search of a linear electrooptic amplitude modulator. *IEEE Photonic. Tech. Lett.* **1991**, *3*, 144–146. [CrossRef]
19. Webster, M.A.; Austin, M.W.; Winnall, S.T. Balanced-bridge Mach-Zehnder Interferometric Optical Modulator with an Electrical Bandwidth of 30Ghz. In Proceedings of the CLEO/Pacific Rim'97 Conference, Chiba, Japan, 14–18 July 1997.
20. Nishihara, H.; Haruna, M.; Suhara, T. *Optical Integrated Circuits*, 1st ed.; McGraw-Hill Book Company: New York, NY, USA, 1985; Chapter 5.
21. *OptiBPM 9.0: Waveguide Optics Design Software*; Optiwave Systems Inc.: Ottawa, ON, Canada, 1989.
22. Hutcheson, L.D. *Integrated Optical Circuits and Components*; Marcel Dekker, INC.: New York, NY, USA, 1987; Chapter 3 (Optical Waveguide Fabrication, p70).
23. Schmidt, R.V. Metal-diffused optical waveguides in LiNbO$_3$. *Appl. Phys. Lett.* **1974**, *25*, 458–460. [CrossRef]
24. Pearsall, T.P.; Chiang, S.; Schmidt, R.V. Study of titanium diffusion in lithium-niobate low-loss optical waveguides by x-ray photoelectron spectroscopy. *J. Appl. Phys.* **1976**, *47*, 4794–4797. [CrossRef]

© 2019 by the author. Licensee MDPI, Basel, Switzerland. This article is an open access article distributed under the terms and conditions of the Creative Commons Attribution (CC BY) license (http://creativecommons.org/licenses/by/4.0/).

Article

Determination of the Chemical Composition of Lithium Niobate Powders

Oswaldo Sánchez-Dena [1,*], Carlos J. Villagómez [1], César D. Fierro-Ruíz [2], Artemio S. Padilla-Robles [1], Rurik Farías [3], Enrique Vigueras-Santiago [4], Susana Hernández-López [4] and Jorge-Alejandro Reyes-Esqueda [1,*]

1. Instituto de Física, Universidad Nacional Autónoma de México, 04510 Mexico City, México
2. Departamento de Mecátronica y Energías Renovables, Universidad Tecnológica de Ciudad Juárez, Avenida Universidad Tecnológica 3051, Colonia Lote Bravo II, 32695 Ciudad Juárez, Chihuahua, México
3. Instituto de Ingeniería y Tecnología, Universidad Autónoma de Ciudad Juárez, Av. Del Charro 450 Norte, 32310 Ciudad Juárez, Chihuahua, México
4. Laboratorio de Investigación y Desarrollo de Materiales Avanzados, Universidad Autónoma del Estado de México, Paseo Colón esquina Paseo Tollocan, 50120 Toluca, Estado de México, México
* Correspondence: ossdena@gmail.com (O.S.-D.); reyes@fisica.unam.mx (J.-A.R.-E.); Tel.: +52-55-5622-5184 (J.-A.R.-E.)

Received: 21 March 2019; Accepted: 30 April 2019; Published: 3 July 2019

Abstract: Existent methods for determining the composition of lithium niobate single crystals are mainly based on their variations due to changes in their electronic structure, which accounts for the fact that most of these methods rely on experimental techniques using light as the probe. Nevertheless, these methods used for single crystals fail in accurately predicting the chemical composition of lithium niobate powders due to strong scattering effects and randomness. In this work, an innovative method for determining the chemical composition of lithium niobate powders, based mainly on the probing of secondary thermodynamic phases by X-ray diffraction analysis and structure refinement, is employed. Its validation is supported by the characterization of several samples synthesized by the standard and inexpensive method of mechanosynthesis. Furthermore, new linear equations are proposed to accurately describe and determine the chemical composition of this type of powdered material. The composition can now be determined by using any of four standard characterization techniques: X-Ray Diffraction (XRD), Raman Spectroscopy (RS), UV-vis Diffuse Reflectance (DR), and Differential Thermal Analysis (DTA). In the case of the existence of a previous equivalent description for single crystals, a brief analysis of the literature is made.

Keywords: chemical composition; lithium niobate; powders; microparticles; nanocrystals

1. Introduction

Nowadays, more than 50 years after Ballman managed to grow large lithium niobate ($LiNbO_3$; LN) crystals with the Czochralski method [1], synthesizing stoichiometric LN single crystals is still a state-of-the-art matter: The reason behind this is the fact that a Z-cut of a stoichiometric grown crystal costs around 12 times more than one possessing a congruent chemical composition [2]. Compared to this version of the material, while comprehensively studied [3] and well exploited technologically [4–6], powders are tacitly considered easier and far less expensive to synthesize. LN powders (LNPws) have served in the past only as survey materials, for example, in the prediction of the nonlinear second order optical capabilities of unavailable single crystals by applying the Kurtz-Perry method in the powdered version [7,8]. Nevertheless, recent developments in LNPws are certainly attracting the attention of scientists and engineers who seek to exploit their potential use in a wide range of applications that span from the construction industry to nonlinear optics.

Cementation materials based on LN have been proposed as a way to improve the air quality of the environment by Artificial Photosynthesis; this is considered important for the reduction of global warming [9]. Regarding LNPws, we emphasize that not only would they be easier than single crystals to implement into cemented materials, but they would also probably enhance the intrinsic surface effects, which are the basis for an improvement of the lifetime of the carriers (photo-generated electrons and holes) involved in Artificial Photosynthesis [10]. Fe-doped LNPws also show, after a post-thermal treatment in a controlled reducing atmosphere, a rather strong ferromagnetic response at room temperature for a doping concentration of the order of 1% mol; this may be considered a first report of the manifestation of ferromagnetism in nanocrystalline LNPws within the regime of very low doping concentrations [11]. Yet in another application based on the powder-in-tube method, a novel fabrication process has been demonstrated for the realization of polarization-maintaining optical fibers [12]. Comprehension of the main mechanism behind this technology, and by looking at the LN mechanical properties [3], it can easily be seen that LNPws are, in principle, good candidates for the fabrication of this type of optical components. Also, possible tuning on the intensity of the Second Harmonic Generation (SHG) that arises from LN micro powders could be ascribed to a proper control of their chemical composition and grain size [13]. This could soon translate into major technical benefits given that neither a critical adjustment of the orientation or temperature in the material (phase matching condition) nor the accurate engineering of a microstructure (quasi-phase matching condition) are substantially needed when the SHG from disordered materials—such as LNPws—are exploited [14].

The performance of LNPws for most of their potentially attributable properties are expected to drastically depend upon their chemical composition (CC), like in the case of single-crystalline LN [3]. Indeed, it has been already demonstrated that size at the nanoscale does not affect the structural symmetry of single LN crystals and that nanosized LN single crystals (down to 5 nm) inherit the nonlinear optical properties from that of large or bulk single crystals [15]: Both the magnitude and the orientation nature of the nonlinear coefficient d_{mn} are preserved. Our work arises from noticing that at least one of the two linear equations that describe the CC of LN single crystals by polarized Raman Spectroscopy measurements [16,17] is not accurate for the case of powders. Hence, it is necessary to properly characterize LNPws, starting by unambiguously determining their CC. Most of the reports found in the literature are only devoted to LN single crystals, where optical and non-optical methods can be found [16–18]. Some of the non-optical methods might also be applied to powders; however, in some cases they would not be accessible to everyone, like neutron diffraction methods, and might also give rise to discrepancies like in cases determining the LN CC by measuring the Curie temperature T_C. Since temperature is a scalar quantity (light propagates and interacts with matter in vector-like form), it would be permissible to expect a single description of the LN CC in terms of T_C that serves for both large single crystals and powders. Interestingly, this is not the case: the systematic measurement of lower T_C values (about 10 °C) for LNPws compared to equivalent single crystals has already been addressed and the reason behind this remains unexplained [18].

In this investigation, a custom-made Raman system has been crafted to obtain control on the polarization state of the light at the excitation and detection stages. With this system, verification of the linear equation for the Raman active mode centered at 876 cm^{-1}, as given by Schlarb et al. [16] and Malovichko et al. [17], can be done on stoichiometric (ST) and congruent (CG) lithium niobate single-crystal wafers, according to the provider [2]. Likewise, this serves to calibrate this assembled system and to confidently state that the aforementioned linear equation does not describe LNPws. Then, with a commercially available system, we observed that the linear relationship remains between the CC of LNPws and the linewidth (Γ) of the same Raman mode (876 cm^{-1}), under which in simpler circumstances the polarization state of light at the excitation and detection stages would be disregarded. In accordance to References [16–19], the accurate determination of the CC of LNPws is proposed by means of a linear fit in terms of the calculated Γ from *non-polarized* Raman spectra. Yet, the main contribution of this work is based on an a priori probing of the formed phases from 11 different synthesized samples by analysis of X-ray diffraction (XRD) experimental data, while relying on the existent information in the phase diagram that describes the pure LN phase along with its surrounding secondary phases (Figure 1). In this way the linear relationship obtained for the averaged Nb content in the crystallites $\langle c_{Nb} \rangle$, in terms of Γ, is affixed to two known or expected values of $\langle c_{Nb} \rangle$ for the two edges that delimit the pure ferroelectric phase: The boundary with phase $LiNb_3O_8$ on one side (Nb excess) and the boundary with phase Li_3NbO_4 on the other (Li excess).

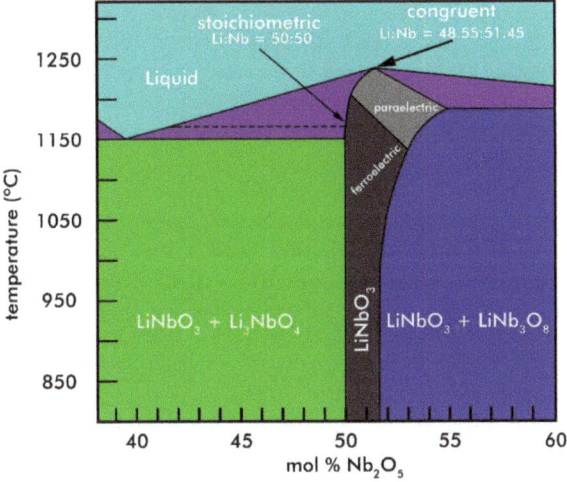

Figure 1. Schematic phase diagram of the Li_2O-Nb_2O_5 pseudo binary system in the vicinity of $LiNbO_3$—redrawn from the publications by Volk and Wöhlecke [3] and Hatano et al. [20].

The nanocrystalline LNPws are obtained by a mechanochemical-calcination route [21,22]. Gradual addition of Li or Nb has been systematically performed by increasing the mass percentage of a precursor containing the desired ion species. Quantification of secondary-phase percentages is carried out with structure refinement by a standard Rietveld method. An alternative linear equation to determine the CC is also given in terms of the calculated cell volumes by means of the same structure refinement. Additionally, linear fitting of the measured band gap energy (E_g), by means of UV-vis Diffuse Reflectance (DR), is also used for this purpose. Differential Thermal Analysis (DTA) is utilized as a verification technique for specific samples and a fourth empirical equation that describes the CC in terms of the Curie temperature is obtained this way. Scanning Electron Microscopy (SEM) is utilized to verify that the particle size distributions do not vary drastically from one sample to another.

2. Materials and Methods

2.1. Synthesis

High purity lithium carbonate (Li_2CO_3) and niobium pentoxide (Nb_2O_5), from Alpha Aesar, were used as starting reagents in a 1:1 molar ratio. The respective masses of the precursors were determined such that 1 g of lithium niobate ($LiNbO_3$; LN) was produced from the following balanced chemical equation:

$$Li_2CO_3 + Nb_2O_5 \rightarrow 2LiNbO_3 + CO_2 \qquad (1)$$

The resultant product was labeled—and hereafter referred to—as LN-STm (ST: stoichiometric, m: mixture) because, in principle, a LN mixture was obtained after milling with a 1:1 molar ratio in terms of Li and Nb. Variations in the chemical composition (CC) of the final resultant powders were sought by adding, at the milling stage, 1–5% of the mass in one of the precursors (with steps of 1% with resolution of 10^{-4} g) while keeping the mass of the other precursor constant, in both cases with respect to the masses measured for sample LN-STm (see Appendix A for table). In this way, 10 more samples were synthesized and labeled as LN + 1%LiP, LN + 1%NbP, LN + 2%LiP, and so on up to LN + 5%NbP (P stands for precursor). It must be clarified that the percentages that appear on these labels are not in terms of the ion species solely, but in terms of the whole mass of the precursors that contain them.

The high-energy milling was carried out in an MSK-SFM-3 mill (MTI Corporation) using nylon vials with YSZ balls; a powder:ball ratio of 0.1 was used for each sample preparation. The milling was performed in 30 min cycles, with 30 min pauses to avoid excessive heat inside the milling chamber, until 200 min of effective milling time was reached. Calcination of the resultant materials (amorphous) was done with a Thermo Scientific F21135 furnace in an air atmosphere. All samples were simultaneously calcined with the following programmed routine: 10 °C/min → 600 °C for 30 min → 2 °C/min → 850 °C for 120 min → cooling down slowly to room temperature.

2.2. X-Ray Diffraction

These patterns were measured in air at room temperature using a Bruker D-8 Advance diffractometer with the Bragg-Brentano θ-θ geometry, a source of CuKα radiation (λ = 1.5406 Å), a Ni 0.5% CuKβ filter in the secondary beam, and a 1-dimensional position sensitive silicon strip detector (Bruker, Linxeye, Karlsruhe, Germany). The diffraction intensity, as a function of the 2θ angle, was measured between 5.00° and 110.00°, with a step of 0.02° every 38.4 s. Sample LN-STm displays a pure ferroelectric lithium niobate (LN) phase, with Bragg peaks resembling those of the COD-2101175 card previously deposited with the Crystallographic Open Database; supplementary crystallographic data can be obtained free of charge from the Web page of the database [23].

Rietveld refinement was performed using computational package X'Pert HighScore Plus from PANalytical, version 2.2b (2.2.2), released in 2006 [24]. Instructions in the section named *Automatic Rietveld Refinement* from the HighScore Online Plus Help document were first followed and then adapted for phase quantification of the samples. In short, an archive with information about the atomic coordinates of LN ("2101175.cif") was downloaded from the Crystallopgraphic Open Database [23]. For the secondary phases $LiNb_3O_8$ and Li_3NbO_4, ICSD-2921 and ICSD-75264 from The Inorganic Crystal Structure Database were used, respectively [25]. The archives were then inserted, along with the experimental data, and Rietveld analysis in "Automatic Mode" was executed, followed by iterative executions in "Semi-automatic Mode," in which different "Profile Parameters" were allowed to vary until satisfactory indexes of agreement were obtained. The averaged crystallite size was also calculated by Rietveld refinement, following instructions from the *Size/Strain Analysis* section; a single lanthanum hexaboride (LaB_6) crystal was used in this case as the standard sample, analyzed with the ICSD-194636 card.

2.3. Raman Spectroscopy

Two Raman systems were employed in this investigation: One custom-made and one of standard use and commercially available. The former allowed for the set-up of different experimental conditions in terms of the polarization state of light at the incident-on-sample and detection stages, including non-polarized, parallel polarized (p), and cross polarized (s) situations. Adopting the so-called Porto's formalism, these experimental conditions were $Z(--)\overline{Z}$, $Z(YY)\overline{Z}$ and $Z(YX)\overline{Z}$, respectively; where, in general, $A(BC)D$ stands for light propagating in the A direction with linear polarization B, before the sample, while selective detection is done on the D direction with polarization C [26].

The commercially available system only featured the non-polarized configuration. It was a Witec alpha300R Confocal-Raman microscope with a 532 nm source of excitation wavelength and 4–5 cm^{-1} of spectral resolution. With this equipment, the Raman spectra were collected in the range 100–1200 cm^{-1} at room temperature and light incident on the normal component of the sample with a power of 3.4 mW; a Nikon 10 objective was used to focus the incoming light on a 1:5 mm spot. An intensity of approximately 11Wcm^{-2} was delivered to the sample. The customized open-air Raman system consisted of an excitation beam output of a continuous wave diode laser at 638 nm wavelength with a power of 37 mW (Innovative photonic solution). The beam was linearly polarized from variable angle mounting and transmitted through a beam splitter to focus the excitation beam into the sample by an aspherized achromatic lens (NA = 0.5, Edmund optics). The excitation spot diameter measured at the focus point had a ~10 µm radius. The collected Raman scattered light from the sample through the aspheric lens and the beam splitter was focused by two silver coated mirrors and one bi-convex lens into a fiber Raman Stokes probe (InPhotonics) that was connected to a QE65 Raman Pro spectrometer (Ocean optics) for a Raman shift range detection between 250–3000 cm^{-1}. In its use for the characterization of the powders, the light at λ = 638 nm was incident at razing angle with P = 10 mW. The Raman spectra were collected in the range 200–1200 cm^{-1} at room temperature with a spectral resolution of 8 cm^{-1}. In this case, a laser intensity of approximately 3 kWcm^{-2} was delivered to the sample. Due to technical issues, most of the utilized experimental conditions were different from one Raman system to another—it is shown how this did not alter the obtained results, except for the detection mode which in both cases was fixed at the backscattering-detection mode (Figure 2).

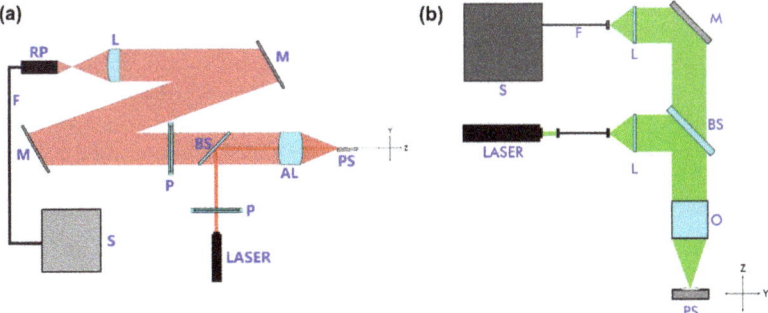

Figure 2. Scheme of the experimental configurations used for the acquisition of Raman spectra: (**a**) Custom-made featuring both configurations, polarized and non-polarized; (**b**) commercially available featuring only non-polarized measurements. From left to right: RP—Raman probe, F—filter, M—mirror, L—lens, P—polarizer, BS—beam splitter, AL—aspheric lens, PS—powdered sample, S—spectrometer, O—objective.

2.4. UV-Vis Diffuse Reflectances and Differential Thermal Analysis

An Ocean Optics USB2000+ UV-VIS Spectrometer and an R400-Angle-Vis Reflection probe were used to collect the diffuse reflectance (DR) spectra of the samples and an Ocean Optics DH-2000-BAL Deuterium-Halogen light source was utilized. Commercially available aluminum oxide (Al_2O_3) was chosen as the standard reference. Precautions were taken so that the approximations necessary to apply the Kubelka-Munk Theory were accomplished [27–29]. These approximations are, mainly speaking, a preparation of the sample being thick enough so that the measured reflectance does not change with further increasing of this parameter (avoidance of Fresnel reflection) and an averaged size of the particles being smaller than such thickness, but larger relative to the wavelength (scattering independent of the wavelength).

The first of these experimental conditions was fulfilled by using a self-supporting pressed powder rectangular mount (3 × 3 × 3 mm); in all the experiments, an amount of approximately 1 g of powder was deposited. The second requirement was fulfilled by determination of the average size particle in the powders, using a field emission Scanning Electron Microscopy (SEM), with a JEOL JSM 5600-LV microscope (V = 20 kV, at 1500×, Mitaka, Tokyo, Japan). The micrographs were analyzed with *ImageJ* software: The edge length histograms were obtained from statistical analysis of at least 200 particles. Lastly, we followed the recommendation of grinding the powders in an agate mortar for a few minutes to avoid sample heterogeneity and regular reflection [29]: All samples were ground for 10 min before measurements.

On the other hand, the Curie temperatures for the samples LN-STm, LN + 1%NbP, LN + 2%NbP, and LN + 3%NbP were measured using differential scanning calorimetry (DSC) equipment coupled to thermogravimetry (TGA), SDT Q600 of TA instruments. The calorimeter was calibrated with respect to the copper melting point (1084 °C). The samples were analyzed in a wide temperature range between room temperature and 1200 °C, at a heating rate of 20 °C/min under a nitrogen atmosphere and using alumina containers. The ferroelectric-paraelectric state transition was observed around 1050–1080 °C. Subsequently, the samples were analyzed in four cooling cycles from 500 °C to 1200 °C at the same heating rate, 20 °C/min, and the process was seen to be reproducible, indicating that there was no permanent change in the volume of the pseudo-ilmenites.

3. Results and Discussion

3.1. X-Ray Diffraction

The obtained XRD pattern for sample LN-STm is shown in the bottom line of Figure 3a. The corresponding pattern of single crystalline LN is in agreement with the one indexed in COD-2101175 [23]. The difference, for all samples, between the obtained XRD patterns (I_{exp}) and their respective calculated patterns by means of Rietveld refinement (I_{ref}) is also shown in the upper half of this figure; for the secondary phases $LiNb_3O_8$ and Li_3NbO_4, ICSD-2921 and ICSD-75264 from The Inorganic Crystal Structure Database were used, respectively [25]. For all cases, this difference function tends to a common baseline, so that neither the formation of thermodynamically stable phases (other than $LiNbO_3$, Li_3NbO_4, and $LiNb_3O_8$) nor the presence of one of the precursors in an interstitial fashion can be deduced, that is, without participating in the formation of one of the involved phases. As seen in this figure, most of the synthesized powders resulted in a pure ferroelectric LN phase, except for samples LN + 4%NbP and LN + 5%NbP (blue lines). Figure 4 and Table 1 have been added for a better visualization of this argument. A loss of Li equivalent to the loss of 5 mol % Li_2CO_3 could be hastily addressed for the central sample LN-STm due to the calcination process. Nevertheless, this information can also be interpreted as having no loss of Li and thus the assumption of a non-ideal sensitivity for the XRD technique must be taken. In other words, a detection threshold of 5.0 mol % Li_2CO_3 = 1.4 mol % Nb_2O_5 exists for 'seeing' a secondary phase by the XRD analysis, combined with the structure refinement, done in this investigation. This assumption has been taken into account in this investigation, thus defining the boundaries that delimit the pure ferroelectric LN phase for samples LN-STm (Li excess) and LN+3%NbP (Nb excess). For the calculation of mol % equivalence between precursors, the values for the relative atomic masses of Li and Nb have been used as presented in the Periodic Table provided by the Royal Society of Chemistry [30].

The calculated cell volumes are plotted in Figure 3b, as a function of the averaged Nb content in the crystallites $\langle c_{Nb} \rangle$, as calculated by the previous procedure (re-labeling of the samples in terms of their predicted CC). A clear linear trend exists for a CC range of 49.7–52.1 mol % Nb_2O_5. Hence, for future reference, we first propose the determination of $\langle c_{Nb} \rangle$ for LNPws in this CC range with the following equation:

$$\langle c_{Nb} \rangle = (8.6207 V_{cell} - 2692.5216) mol\ \%\ \pm 0.5\ mol\ \% \quad (2)$$

where V_{cell} stands for the cell volume in (angstrom)^3 units, calculated by a standard structure refinement method. The 0.5 mol % uncertainty is determined by the sum of the uncertainty associated to the linear fitting (0.14 mol % Nb_2O_5) and half the longer step in the Nb precursor (0.53/2 = 0.27 mol % Nb_2O_5), both multiplied by the square root of the averaged goodness of fit factor for the six involved samples ($\sqrt{1.55}$). The uncertainty associated with the linear fitting has been determined following several calculations according to Baird [31].

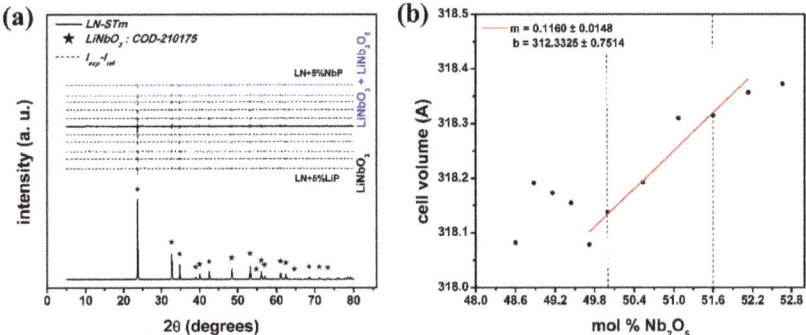

Figure 3. XRD results: (**a**) Experimental pattern of sample LN-STm and, for all samples, the differences between experimental and their respective calculated patterns with Rietveld refinement. The central sample, LN-STm, is distinguished from the rest by the solid line; (**b**) cell volume as a function of mol % Nb precursor. The edges of the ferroelectric pure LN phase are represented by the vertical dashed lines.

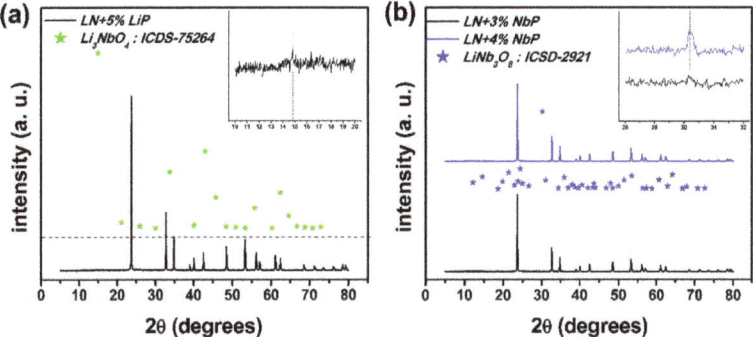

Figure 4. X-ray diffraction patterns close to the boundaries of the pure ferroelectric LN phase: (**a**) Under the assumption of no loss of Li, sample LN-STm is on the excess of the Li boundary; (**b**) sample LN + 3%NbP is on the excess of Nb boundary.

Table 1. Phase percentages present in the synthesized samples, along with the calculated cell volumes and relevant agreement indices of the refinement process.

Sample	% LiNbO3	% Li3NbO4	% LiNb3O8	Cell Volume ($\mathring{A}^3$)	Weighted R Profile	Goodness of Fit
LN+5%LiP	99.9	0.1	0	318.0820	5.82	2.03
LN+4%LiP	100	0	0	318.1917	5.24	1.48
LN+3%LiP	100	0	0	318.1732	5.58	1.50
LN+2%LiP	100	0	0	318.1546	5.60	1.49
LN+1%LiP	100	0	0	318.0787	5.70	1.52
LN-STm	100	0	0	318.1374	5.71	1.57
LN+1%NbP	100	0	0	318.1930	5.52	1.55
LN+2%NbP	100	0	0	318.3095	5.71	1.53
LN+3%NbP	100	0	0	318.3149	5.54	1.65
LN+4%NbP	98.2	0	1.8	318.3566	5.54	1.51
LN+5%NbP	97.8	0	2.2	318.2735	5.54	1.57

Justification of the Assumption made in the X-Ray Diffraction Analysis

The reasoning behind the assumption made can be summarized in three main points. First, a good agreement can be seen with the phase diagram (Figure 1), upon which by close inspection, around T = 850 °C, a CC range of approximately 1.7 mol % Nb_2O_5 is deduced for the pure ferroelectric LN phase. In this investigation, the observed range goes from the ST point $\langle c_{Nb}\rangle$ = 50.0 mol % (sample LN-STm) to a near-CG point $\langle c_{Nb}\rangle$ = 53.0 − 1.4 = 51.6 mol % (sample LN + 3%NbP), that is Δc_{pureLN} =1.6 mol % Nb_2O_5. A direct explanation would not be found for an estimated range of 4.4 mol % Nb_2O_5 if this assumption had not been taken. Secondly, under these circumstances it follows that, out of 11 synthesized samples, only samples LN-STm, LN + 1%NbP, LN + 2%NbP, and LN+3%NbP resulted to have a pure ferroelectric LN phase. It will be soon shown that, for all the performed studies, unmistakable linear relationships happen to exist among these samples and their corresponding experimental parameters (related to the CC); a striking, very sensitive, deviation from these trends is observed for all samples out of this range, in some cases even under the consideration only of neighbor samples such as LN + 1%LiP and LN + 4%NbP. Lastly, besides the well-known difficulties to produce single-phase ST LN at temperatures used in solid-state reactions (T ≥ 1200 C) [32,33], much ambiguity can be found in the literature concerning deviation from stoichiometry in the formation of LNPws at calcination temperatures near T = 850 °C. While only one work is found to report no loss of Li after two 16-hour reaction periods at 1120 °C [34], other authors have observed the loss of Li at 600–800 °C within at least three different investigations [33,35,36]. However, these methods of synthesis are very different from each other, except for those in the works published in 2006 (Liu et al.) [33] and 2008 (Liu et al.) [36], which are aqueous soft-chemistry methods. The deviation from stoichiometry tendency in the formation of LNPws through aqueous soft-chemistry methods, in comparison to non-aqueous (as in this investigation), has already been identified [37]. Besides, high-energy milling has previously been proposed as a method to prevent loss of Li, in contrast to Pechini's method, sol-gel, and coprecipitation [21].

It is also worth mentioning that De Figueiredo et al. [38] had a similar observation in their investigation: They had a small amount of non-reacted Li_2CO_3 not detected by XRD, but only identified after DTA and Infrared Spectroscopy; the LNPws were synthesized via mechanical alloying. They explained this observation by assuming that the number and size of the Li_2CO_3 nanocrystals were sufficiently low and small to not being detected by XRD. Hence, the assumption taken of no loss of Li and the existence of a detection threshold of 1.4 mol % Nb_2O_5 in XRD might have been justified with these lines. This detection threshold can be considered unique and expected to change according to different experimental variables and analysis tools, including spatial and temporal size of the step during the experiment, brand, and model of equipment utilized, as well as the software used for Rietveld refinement, among others.

3.2. Raman Spectroscopy

Verification of the linear equation for the Raman active mode centered at 876 cm^{-1} [16,17] was done by using the assembled Raman system (Figure 2a) on the aforementioned stoichiometric (ST) and congruent (CG) lithium niobate (LN) wafers. Even though the experimental conditions therein described were not exactly reproduced, this could be accomplished within the given absolute accuracy and, thus, calibration of this equipment could be done. At this instance, use of the equation for the Raman band located at 876 cm^{-1} has been done [16,17]. A detailed description of the phonon branches of single crystal LN and their assignment can be found elsewhere [39,40]. No specifications regarding the resolution of the Raman bands or fitting techniques are given by Schlarb et al. [16] or Malovichko et al. [17], although these procedures are critical for achieving great accuracy in the determination of the LN CC [16,39–41]. Moreover, it is not clearly stated whether the complete linewidth (Γ), or just the halfwidth, is to be entered in this equation.

The resolution of this Raman band was explored, after normalization of the full spectra, by two distinct line shape fittings: Gaussian and Lorentzian. The Full Width at Half Maximum (FWHM; Γ)

was extracted from the fitting (*Origin Pro 8*) and used in the calculations. Change of the intercept value from 53.29 to 54.8 had also been tried, as suggested whenever no polished single crystals are available [16]. From all the calculations performed, we noticed that only for those (halfwidths) under a Lorentzian fit and using the intercept value of 54.8, the calculated Li contents follow this equation within the uncertainty of 0.2% mol, which "govern the absolute accuracy of the described method" [16]. The values obtained by this calculation were $\langle c_{Li} \rangle$ =50.3 mol % for the ST wafer and 48.5 mol % for the CG one. Thus, this approach has been adopted for the investigation with the LNPws. Before presenting these results, one more point needs to be further discussed.

It can be argued that the value of 50.3 mol % for the ST wafer goes out of the uncertainty range, thus not justifying the implications made above. Nevertheless, it must be noticed that the experimental conditions used in this investigation are subtly different from those described by Schlarb et al. [16] and Malovichko et al. [17]. Succinctly, they used an experimental $Z(YY)X$ configuration (using Porto's convention [26]), whereas for our case, given certain technical limitations, a $Z(YY)\overline{Z}$ configuration was used in this investigation. Besides, no direct statement concerning the propagation of light along an axis of the crystals studied is done by these authors, but it can be inferred that they excited along the crystallographic Z-axis by recalling the condition of zero (or small) phonon directional dispersion to simplify their adjustments (band resolution) [16]. In our case, wafers with Z-cuts were used, upon which light was made to impinge on normal to the surface. The incident radiation then propagates in a plane containing the extraordinary axis, inducing in this way short-range atomic forces (extraordinary refractive index) that compete to long-range atomic forces behind the splitting of longitudinal optic (LO) and transverse optic (TO) phonons [42]. Significant changes in the Raman spectra of LN single crystals, especially in the position of the bands located at 153 cm^{-1} and 578 cm^{-1} (red and blue shifts), have already been identified and addressed to the overlapping of the LO and TO lattice vibrations [42–44]. Such an overlapping is clearly a drawback for band resolution and it might be the reason behind the discrepancy between predicted and measured values; interestingly, this is only relevant in single crystals of ST composition.

Application of the same procedure to the synthesized LNPws gives unsatisfactory results, according to the implications obtained from the XRD analysis (re-labeling of the samples in terms of their predicted CC, Figure 3b). As expected, the same occurs if this is applied to the non-polarized Raman spectra. It worsens considering the Raman band is located at 153 cm^{-1}, where the corresponding linear equation is used, and the Raman spectra are measured with the commercially available Raman system (Witec), which features recording of intensity in the range 0–200 cm^{-1}. However, well defined linear trends can be seen for the calculated Raman halfwidths around the pure LN ferroelectric phase, but only for the case of the band at 876 cm^{-1} as measured under non-polarized experimental conditions. For both situations (Witec and self-assembled systems), the trend is of an increasing halfwidth with decreasing Li content; surprisingly, despite the great differences between both experimental configurations and conditions (Figure 2), both trends are very similar. This feature can also be seen for the positions of the bands (x_c), and it remains for the resultant values of the halfwidths divided by the positions ($\Gamma/2x_c$). Figure 5b shows how this $\Gamma/2x_c$ parameter relates to the Nb content of the synthesized powders, as determined by XRD analysis. Given the similarity between the results obtained by both experimental configurations, this graph represents the average of such results. For sample LN-STm, the Raman spectra measured with the Witec system are shown in Figure 5a; these closely resemble those obtained in polycrystalline LN by Repelin et al. [40].

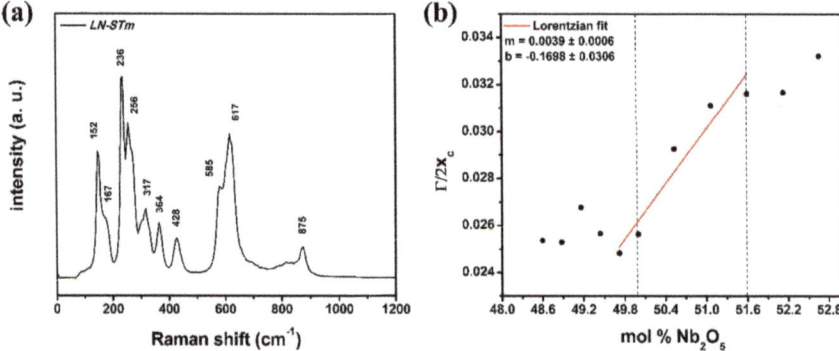

Figure 5. Results obtained by Raman Spectroscopy: (**a**) Non-polarized Raman spectra of the central sample LN-STm, obtained with the commercially available Raman system; (**b**) Linear trend upon which Equation (3) is based for the case of band resolution with a Lorentzian fit, averaged calculated data from those obtained by two distinct Raman systems.

As the resolution of this Raman band (876 cm^{-1}) by means of a Gaussian fitting does not entail significant changes either, the following equations are proposed for the determination of $\langle c_{Nb} \rangle$ in LNPws:

$$\langle c_{Nb} \rangle_L = \left(256.4103 * \left(\frac{\Gamma_L}{2x_c}\right) + 43.5385\right) mol\ \%\ \pm 0.4\ mol\ \%$$
$$\langle c_{Nb} \rangle_G = \left(588.2353 * \left(\frac{\Gamma_G}{2x_c}\right) + 42.7059\right) mol\ \%\ \pm 0.5\ mol\ \%$$
(3)

where Γ_i stands for the FWHM in cm^{-1} of the Raman band around 876 cm^{-1}, resolved by linear fitting either using a Lorentzian or a Gaussian line shape, x_c denotes the center of this Raman band. Normalization of the full Raman spectra precedes the linear fitting and, regardless of the line shape, enlargement around this band is suggested, extending it as much as possible (precise determination of the baseline) and applying a single or double-peak fitting, rather than performing a multi-peak fitting to the full Raman spectra. Like in the XRD analysis, the uncertainty is determined by summation over half the longer step in the Nb precursor (0.53/2 = 0.27 mol % Nb$_2$O$_5$), the uncertainty associated to the linear fitting (0.12 mol % Nb$_2$O$_5$ (0.23 mol % Nb$_2$O$_5$) for the Lorentzian (Gaussian) case), and dividing by the square root of the averaged (five involved samples) reduced χ^2 fit factor obtained in the resolution of the band $\sqrt{0.9823}$ ($\sqrt{0.9866}$). Once more, the uncertainty associated to the linear fitting is determined following several calculations according to Baird [31].

Lastly, the fact that the trend remains linear is not surprising. Scott and Burns [34] have previously demonstrated this, based on experimentation; showing in this way that the Raman spectra from poly-crystalline LN inherits the essential features of those from single crystal LN [45]. Conceptually, this can be understood by recalling the intrinsic nature of LN to deviate from the stoichiometric point. Under regular circumstances, LN contains high amounts of intrinsic defects such as anti-site Nb ions (Nb_{Li}), which are compensated by their charge-compensating Li vacancies (V_{Li}) [3,46]. Such a substitution mechanism imposes fundamental changes on the electronic structure, inducing in this way, variations in the macroscopic dielectric tensor of LN [16]. Yet, because in this substitution mechanism gradual Nb increments are proportional to the decreasing of Li, the variations of the dielectric tensor are expected to be linear, as far as the Nb-Li interchange is sufficiently small.

3.3. UV-Vis Diffuse Reflectances and Differential Thermal Analysis

The sensitivity of the chemical composition (CC) of lithium niobate (LN) to the fundamental band gap or fundamental absorption edge has been previously reported for LN single crystals [47,48]. Kovács et al. have given a corresponding linear equation with different sets of fitting parameters, depending on the character of the refractive index (ordinary and extraordinary), and the definition of the absorption edge (either as corresponding to a value in the absorption coefficient of 20 cm^{-1} or 15 cm^{-1}) [48]. There is no point in using this equation to describe the CC of LNPws, since these terms (refractive index and absorption coefficient) make no sense when related to powders.

In this investigation, the direct measurements of the DR spectra for the 11 samples are transformed to the Kubelka-Munk (K-M) or *remission* function $F(R_\infty)$, straightforwardly with the acquisition software (*Spectra Suite*). Since this function is a proxy of the actual absorption spectrum [29], these data are used to find the fundamental absorption edge for all the samples. For practical purposes, a direct band gap is assumed for LN—notice that it could also be assumed to be indirect [49]. Thus, under this assumption, the fundamental band gap is proportional to the square of the remission function, as is shown in Figure 6a,b. The Nb content of LNPws is linearly related to the fundamental band gap E_g (Figure 6c). Equation (4) allows us to accurately determine the Nb content of a determined sample, in terms of E_g (in eV units).

$$\langle c_{Nb} \rangle = (3.9078 * E_g + 34.6229) mol\% \pm 0.4\ mol\% \tag{4}$$

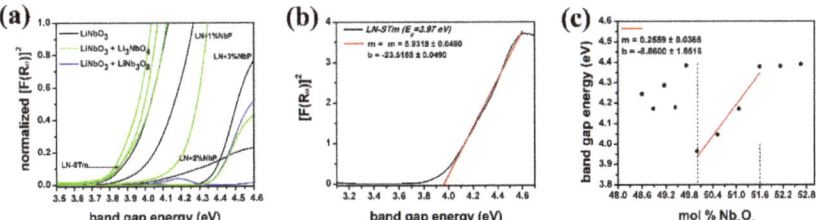

Figure 6. Graphics derived from analysis of the data obtained by UV-vis Diffuse Reflectance measurements: (**a**) Normalized Kubelka-Munk or remission functions in terms of the energy of the light in eV units; (**b**) Demonstration of the determination of the onset for sample LN-STm (assuming a direct interband transition) to determine the fundamental band gap energy; (**c**) Fundamental band gap energy as a function of mol % Nb precursor.

Interaction of light with matter at a fundamental level must be considered in the DR and Raman Spectroscopy techniques. In other words, because of the ubiquitous randomness of the media, strong scattering effects are present in both Rayleigh (crystallite size) and Mie scattering (particle size). The study of the intensity and angular distribution of the scattered field by the powders has not been done on this investigation; however, certainty of the results obtained by these techniques is expected under certain limits if no large variations in the crystallite and particle average sizes are found. Considering all the synthesized samples, the resultant average crystallites are distributed in a 100–300 nm range, with overall mean and standard deviation values of 157 and 58 nm, respectively. Also, for four randomly chosen samples, the distributions in particle size were determined by statistical analysis of micrographs obtained by Scanning Electron Microscopy (SEM). The obtained distributions were very similar and the centers (x_c) of these distributions fall within a band 1 µm thick, centered at 2.6 µm, as shown in Figure 7.

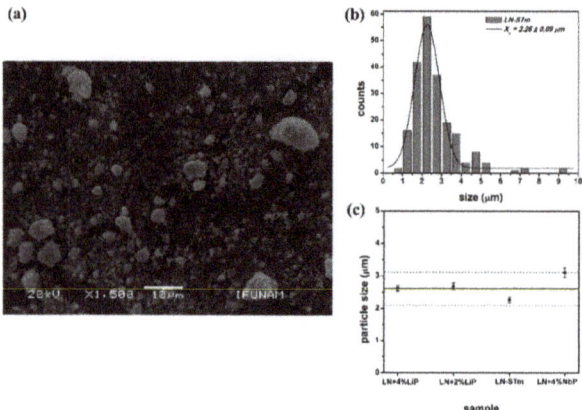

Figure 7. Information derived from SEM: (**a**) and (**b**) Micrograph and particle size distribution for sample LN-STm, respectively; (**c**) Centers of the particle size distributions obtained for four randomly-chosen samples.

Regarding the ferroelectric-paraelectric phase transition, a change in the crystalline structure of LN occurs in which the symmetry of the system increases [3]. This second-order phase transition is described by the Landau order-disorder theory, where a finite discontinuity in the heat capacity of the system having this transition has been addressed as a direct thermodynamic consequence [50]. Figure 8a shows the difference in temperature between the reference container for each of the studied samples; with this technique, only samples presenting a pure LN ferroelectric phase have been investigated. The Curie temperatures are determined by the extrapolated departure from the baseline, these being plotted in Figure 8b in terms of the Nb content. A nonlinear quadratic trend better describes this relation, with fitting coefficients A = 18623.560, B = -667.969, and C = 6.383; as expected, this is also the case for LN single crystals [51]. Nevertheless, use of the linear fitting coefficients is done in the analysis that follows, so that a simple calculation of an uncertainty value follows by use of Equation (5), where the Curie temperature T_C, is in Celsius.

$$\langle c_{Nb} \rangle = (-0.0515 * T_c + 110.8505)\,mol\,\% \;\pm 0.4\;mol\,\% \tag{5}$$

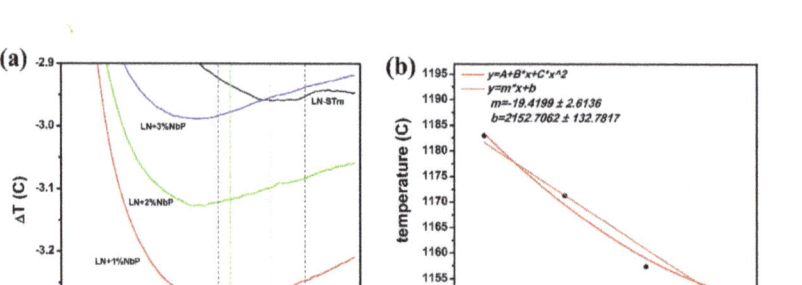

Figure 8. Thermometric results: (**a**) Curie temperatures as a function of mol % Nb precursor; (**b**) Obtained curves for samples within the pure LN phase. The Curie temperatures are determined by extrapolation of the departure from the baseline.

Use of this equation gives an estimate of T_C = 1181.56 °C (1153.41 °C for a ST (CG) powder); whereas, with the quadratic expression, it is of 1182.61 °C and 1153.01 °C, respectively. These values vary in no more than 0.1%. Regarding single crystals, a variation of 0.7% can be found for the Curie temperatures calculated for these CCs, by use of equations reported in two independent investigations [51,52]. Using the equation given by Bordui et al. [52], the calculated values are T_C = 1206.47 °C (1149.83 °C) for the ST (CG) crystal. Thus, contrary to what was believed, not a unique description of the LN CC regardless of its version (powder or single crystal) can be formulated by DTA either. This observation of the T_C being lower for ST LNPws, with respect to ST LN crystals, has been previously noticed [36], apart from the observations highlighted in the introduction. A straight explanation of this subtlety cannot be found nowadays in the literature. A classic theoretical development shows that the energy of the vibrations within the structure is the dominant contribution to the heat capacity—if the elastic response of a crystal is a linear function of the applied forces [53]. Thus, it is inferred that this might be explained under consideration of anharmonic crystal interactions, that is, phonon-phonon coupling. Still, further investigations on these matters are needed.

Lastly, it is acknowledged that in contrast to pioneering works (on LN single crystals, References [16] and [17]) the Equations (2)–(5) here give the averaged Nb content in the crystallites $\langle c_{Nb} \rangle$ instead of $\langle c_{Li} \rangle$. Although a simpler comparison with data in the literature could have been attained by putting these equations in terms of $\langle c_{Li} \rangle$, it was decided to do it in terms of $\langle c_{Nb} \rangle$ because of a simpler interpretation and association with a phase diagram describing LN, like that given in Figure 1. It has been noticed that most of the phase diagrams existent in the literature to describe LN, not to say all, are presented in terms of Nb_2O_5 mol %. This is readily understood since even in the fabrication of large LN single crystals, melts of Nb_2O_5 and another compound containing Li are used [3,20]. The equivalent equations in terms of $\langle c_{Li} \rangle$ are given in Appendix B.

3.4. Grinding of a Single Crystal

The bought single crystal with stoichiometric composition, described above, was turned into powder with ST composition. Low-energy grinding with an agate mortar was employed discontinuously in several steps until an averaged particle size of 1.6 μm (checked by SEM) was reached. In some instances, commercial acetone (purity ≥ 99.5%, Sigma-Aldrich) was used to ease the grinding, especially during its initial stages. Verification of Equations (2) and (3) was sought by repeating the experimental procedures performed on the synthesized powders; in the case of Raman Spectroscopy (RS), only the commercially available system (Witec alpha 300R) was used. The results obtained are shown in Table 2. While RS does imply a chemical composition according to what was expected, stoichiometrically (50 mol % Nb_2O_5), the structure refinement does not. This can be attributed to changes of the lattice parameter (lattice distortion) due to a variable local lattice strain frequently observed in nanocrystalline materials, induced by excess of volume at the grain boundaries [54]. Remarkably, our powdered single crystal differs from the synthesized powders in the averaged crystallite size: On the latter, a myriad of nanosized crystals (100–300 nm) form large particles of the order of 2–3 μm (see Figure 7c), while on the former it can be argued that crystallite size equals the particle size; actually, the applied Rietveld refinement for the calculation of the averaged crystallite size of the grinded crystal does not converge. These implications must be confirmed and scrutinized by further investigation. Lastly, since Equation (3) is strongly dependent on the XRD analysis (re-labeling of the samples in terms of their predicted CC), the Raman results shown in Table 2 demonstrate the reliability of our method.

Table 2. Estimated chemical composition for the grinded single crystal.

Experimental Technique	Measured Parameter	Associated Error Parameter	Equation Utilized	Nb Content (mol % Nb_2O_5)
XRD + Rietveld refinement	Cell volume: 317.9234 A°	Goodness of Fit: 1.8756	(2)	48.2
Raman Spectroscopy	Γ/x_c: 45.3038cm^{-1}/873.9676 cm^{-1}	Reduced $\chi^{(2)}$: 4.70 × 10^{-6}	(3), Lorentz fit	50.2
	Γ/x_c: 21.8202cm^{-1}/874.1964 cm^{-1}	Reduced $\chi^{(2)}$: 8.38 × 10^{-6}	(3), Gaussian fit	50.1

4. Conclusions

Despite the increasing interest in lithium niobate powders (LNPws) due to their importance in possible applications in actual and future nanooptoelectronic devices, as well as the facility to produce them in large quantities, an accurate and trusting method to determine their chemical composition (CC) does not exist, to the best of our knowledge. Therefore, in this work a first step is given in this direction by developing a facile method based mainly on imposing X-Ray Diffraction (XRD) as a seed characterization technique. Raman Spectroscopy, UV-vis Diffuse Reflectance and Differential Thermal Analysis enrich this work, representing various alternatives for the independent and accurate determination of the CC of LNPws. An empirical equation that describes this fundamental property in terms of a corresponding experimental parameter is given for each of these four characterization techniques.

We wish to underline here the main aspects of our method. It is primarily based on the quantification of pure and secondary phase percentages by XRD, followed by Rietveld structure refinement. Secondly, relying on the LN phase diagram, the CCs of the studied samples are inferred, and thereafter labeled in terms of the Nb content in the crystallites. Lastly, having done this, any of the mentioned characterization techniques can be used to relate such a labeling with their corresponding experimental parameter. In the case of a user who wants to determinate the CC of LNPws only, he/she would only need to perform the last step and make use of any of Equations (2)–(5), respectively. On the other hand, in the case of wanting to describe other powders apart from LNPws, the whole method (three main steps described above) might be further applied inasmuch as akin materials are to be investigated, lithium tantalate (LiTaO$_3$) powders for example.

The validity of this methodology is proven self-consistently with the determination of the CC of several samples, where the content of Li and Nb is varied in a controlled way. According to a paramount observation made in the peer reviewing process of this article, the main shortcoming of this investigation is the large uncertainty associated with Equations (2)–(5). Rigorously, they should not be used for a practical composition determination and, instead, it only could be stated with more confidence that, by using these equations, the composition of a LN powder would be closer to the stoichiometric or congruent compositions, or rather in an intermediate state. However, both, the resolution and the associated uncertainties of this methodology, can be significantly improved by analyzing larger quantities of powder. As mentioned in the details related to the uncertainty calculations and given after introducing Equations (2) and (3), the major contribution to uncertainty emerges from the determination of the boundaries of the pure ferroelectric LN phase: Determined by dividing the $\Delta c_{pureLN} = 1.6$ mol % Nb$_2$O$_5$ by three increasing steps of Nb content, and then dividing by 2 (0.53/2 = 0.27 mol % Nb$_2$O$_5$). The associated uncertainty to Equations (2)–(5) can be significantly reduced if a larger number of samples are synthesized in this range, which can be more easily achieved if larger quantities of powder are prepared. As an example, it is expected that by synthesizing approximately 10 g of powder, around 40 points would be available for analysis if the increasing step is fixed at 0.1% in the mass of the Nb precursor, resulting in a decrease in the overall uncertainties of about 50–80% (noticing that the uncertainty associated with the linear fitting would also be reduced significantly). Conclusively, although it is acknowledged that the proposed equations are not universal in the sense that they may only describe the CC of LNPws with specific physical properties (crystallite and particle dimensions), this work paves the way to furnish a general description and claims the attention of the community advocated to this field to broaden the present results. For a more general description, besides the synthesis of larger number of samples, the influence of other experimental factors and parameters such as the method of synthesis, the beam spot size, the intensity of light (Raman Spectroscopy), the averaged crystallite and particle size, and randomness, among others, should be considered in future investigations.

No full credit for all the ideas developed in this work is to be taken. The idea of determining the CC of LNPws by means of a linear fit to data obtained from Raman spectra was first conceived in the pioneering work of Scott and Burns in 1972 [34]. Indeed, no equation is given in this work, but it could be easily extracted from Figure 3 (in Reference [34]) to describe LNPws instead of LN single crystals;

again, it would not be easy to decide whether the complete linewidth (Γ), or just the halfwidth, is to be entered in such a hypothetical equation, and if a Lorentzian or Gaussian distribution is to be used. The work of Scott and Burns is also a pioneer to the ideas conceived by Schlarb et al. [16] and Malovichko et al. [17], whom later in 1993 exploited this fruitful result and demonstrated that other optical processes, besides Raman Spectroscopy, resulted into data that fit linearly with the LN CC. Also, regarding Equation (2), the previous observation of an increase of the lattice parameters or cell volume with increasing Nb content is also acknowledged [55]. An equation is formulated in Reference [18] from the data given in [55]. Interestingly, equation (4) in Reference [18] is almost the same as Equation (A1), given in Appendix B, if the slope and intercept values of the latter are divided by a constant value of 2.58; the very small discrepancy might be attributed to variation in the local lattice strain, as discussed above where the results of grinding a LN single crystal of stoichiometric composition are presented. At last, apart from providing four distinct alternatives to describe accurately the CC of LNPws (instead of single crystalline LN), what is innovative in the present work is the self-consistency character of the whole method: no other technique is needed to confirm the CC of the powders since the determination of the pure ferroelectric LN phase boundaries by XRD analysis suffices for this purpose. The four distinct methods are based on standard characterization techniques, accessible nowadays to large scientific communities in developing countries.

Author Contributions: Conceptualization, O.S.-D., C.D.F.-R. and R.F.; methodology, O.S.-D. and C.D.F.-R.; software, O.S.-D.; validation, A.S.P.-R. and S.H.-L.; formal analysis, O.S.-D.; investigation, O.S.-D., C.J.V., C.D.F.-R., E.V.-S. and S.H.-L.; resources, C.J.V., R.F., E.V.-S. and J.-A.R.-E.; writing—original draft preparation, O.S.-D.; writing—review and editing, J.-A.R.-E.; visualization, C.D.F.-R. and A.S.P.R.; supervision, R.F. and J.-A.R.-E.; project administration, J.-A.R.-E.; funding acquisition, J.-A.R.-E.

Funding: This research was funded by CONACyT, and partially funded by PIIF-3-2018 and UNAM-PAPIIT [grant numbers IN112919, IN114317].

Acknowledgments: The authors wish to acknowledge the technical assistance of Antonio Morales Espino and Manuel Aguilar Franco. The assistance from Alicia Torres and Maricruz Rocha (Laboratorio de Análisis de Materiales, UTCJ) with the milling of the samples is also acknowledged. O.S.-D. thanks CONACyT scholarship grant and Laboratorio Universitario de Caracterización Espectroscópica (LUCE), ICAT-UNAM for providing equipment for Raman spectroscopy measurements. The authors also thank the revision of the final English text by Fernando Funakoshi.

Conflicts of Interest: The authors declare no conflict of interest. The funders had no role in the design of the study; in the collection, analyses, or interpretation of data; in the writing of the manuscript, or in the decision to publish the results.

Appendix A

In the following table the measured values for the masses of the precursors used in each of the 11 synthesis procedures are given.

Sample	Nb_2O_5 Mass (g)	Li_2CO_3 Mass (g)	Sample	Nb_2O_5 Mass (g)	Li_2CO_3 Mass (g)
LN+5%LiP	0.8989	0.2622	LN+1%NbP	0.9079	0.2498
LN+4%LiP	0.8988	0.2598	LN+2%NbP	0.9167	0.2496
LN+3%LiP	0.8991	0.2574	LN+3%NbP	0.9259	0.2497
LN+2%LiP	0.8990	0.2547	LN+4%NbP	0.9348	0.2498
LN+1%LiP	0.8989	0.2523	LN+5%NbP	0.9438	0.2498
LN-STm	0.8990	0.2498			

Appendix B

Equations in terms of the averaged Li content in the crystallites $\langle c_{Li} \rangle$ would also be useful, especially when comparing to measurements on single crystals described elsewhere [16–18]. Equations (2)–(5) in terms of $\langle c_{Li} \rangle$ are:

$$\langle c_{Li} \rangle = (-7.6453 V_{cell} + 2482.2171) mol\ \% \quad \pm 0.5\ mol\ \% \tag{A1}$$

$$\langle c_{Li}\rangle_L = \left(-259.0674 * \left(\tfrac{\Gamma_L}{2x_c}\right) + 56.8135\right) mol\% \ \pm 0.4\ mol\%$$
$$\langle c_{Li}\rangle_G = \left(-588.2353 * \left(\tfrac{\Gamma_G}{2x_c}\right) + 58.0412\right) mol\% \ \pm 0.5\ mol\% \tag{A2}$$

$$\langle c_{Li}\rangle = \left(-3.9602 * E_g + 65.5987\right) mol\% \ \pm 0.4\ mol\% \tag{A3}$$

$$\langle c_{Li}\rangle = (0.0519 * T_c - 11.3805) mol\% \ \pm 0.4\ mol\% \tag{A4}$$

References

1. Ballman, A.A. Growth of Piezoelectric and Ferroelectric Materials by the Czochralski Technique. *J. Am. Ceram. Soc.* **1965**, *48*, 112–113. [CrossRef]
2. MTI Corporation, LiNbO3 & Doped. Available online: http://www.mtixtl.com/linbo3.aspx (accessed on 29 January 2019).
3. Volk, T.; Wöhlecke, M. Point Defects in LiNbO₃. In *Springer Series in Materials Science. Lithium Niobate. Defects, Photorefraction and Ferroelectric Switching*, 1rst ed.; Hull, R., Osgood, R.M., Jr., Parisi, J., Warlimont, H., Eds.; Springer: Berlin/Heidelberg, Germany, 2009; Volume 115, pp. 9–50. ISBN 978-3-540-70765-3.
4. Weis, R.S.; Gayklord, T.K. Lithium Niobate. Summary of Physical Properties and Crystal Structure. *Appl. Phys. A* **1985**, *37*, 191–203. [CrossRef]
5. Luo, R.; Jiang, H.; Rogers, S.; Liang, H.; He, Y.; Lin, Q. On-chip second-harmonic generation and broadband parametric down-conversion in a lithium niobate microresonator. *Opt. Exp.* **2017**, *25*, 24531–24539. [CrossRef] [PubMed]
6. Pang, C.; Li, R.; Li, Z.; Dong, N.; Cheng, C.; Nie, W.; Bötger, R.; Zhou, S.; Wang, J.; Chen, F. Lithium Niobate Crystal with Embedded Au Nanoparticles: A New Saturable Absorber for Efficient Mode-Locking of Ultrafast Laser Pulses at 1µm. *Adv. Opt. Mater.* **2018**, *6*, 1800357. [CrossRef]
7. Kurtz, S.K.; Perry, T.T. A Powder Technique for the Evaluation of Nonlinear Optical Materials. *J. Appl. Phys.* **1968**, *39*, 3798–3812. [CrossRef]
8. Aramburu, I.; Ortega, J.; Folcia, C.L.; Etxebarria, J. Second harmonic generation by micropowders: A revision of the Kurtz-Perry method and its practical application. *Appl. Phys. B: Lasers Opt.* **2014**, *116*, 211–233. [CrossRef]
9. Nath, R.K.; Zain, M.F.M.; Kadhum, A.A.H. Artificial Photosynthesis using LiNbO₃ as Photocatalyst for Sustainable and Environmental Friendly Construction and Reduction of Global Warming: A Review. *Catal. Rev. Sci. Eng.* **2013**, *56*, 175–186. [CrossRef]
10. Yang, W.C.; Rodriguez, B.J.; Gruverman, A.; Nemanich, R.J. Polarization-dependent electron affinity of LiNbO₃ surfaces. *Appl. Phys. Lett.* **2004**, *85*, 2316–2318. [CrossRef]
11. Fierro-Ruíz, C.D.; Sánchez-Dena, O.; Cabral-Larquier, E.M.; Elizalde-Galindo, J.T.; Farías, R. Structural and Magnetic Behavior of Oxidized and Reduced Fe Doped LiNbO₃ Powders. *Crystals* **2018**, *8*, 108. [CrossRef]
12. Kudinova, M.; Humbert, G.; Auguste, J.L.; Delaizir, G. Multimaterial polarization maintaining optical fibers fabricated with powder-in-tube technology. *Opt. Mater. Express* **2017**, *10*, 3780–3790. [CrossRef]
13. Sánchez-Dena, O.; García-Ramírez, E.V.; Fierro-Ruíz, C.D.; Vigueras-Santiago, E.; Farías, R.; Reyes-Esqueda, J.A. Effect of size and composition on the second harmonic generation from lithium niobate powders at different excitation wavelengths. *Mater. Res. Express* **2017**, *4*, 035022. [CrossRef]
14. Skipetrov, S.E. Disorder is the new order. *Nature* **2004**, *432*, 285–286. [CrossRef]
15. Knabe, B.; Buse, K.; Assenmacher, W.; Mader, W. Spontaneous polarization in ultrasmall lithium niobate nanocrystals revealed by second harmonic generation. *Phys. Rev. B* **2012**, *86*, 195428. [CrossRef]
16. Schlarb, U.; Klauer, S.; Wesselmann, M.; Betzler, K.; Wöhlecke, M. Determination of the Li/Nb ratio in Lithium Niobate by Means of Birefringence and Raman Measurements. *Appl. Phys. A* **1993**, *56*, 311–315. [CrossRef]
17. Malovichko, G.I.; Grachev, V.G.; Kokanyan, E.P.; Schirmer, O.F.; Betzler, K.; Gather, B.; Jermann, F.; Klauer, S.; Schlarb, U.; Wöhlecke, M. Characterization of stoichiometric LiNbO₃ grown from melts containing K₂O. *Appl. Phys. A: Mater. Sci. Process.* **1993**, *56*, 103–108. [CrossRef]
18. Wöhlecke, M.; Corradi, G.; Betzler, K. Optical methods to characterise the composition and homogeneity of lithium niobate single crystals. *Appl. Phys. B* **1996**, *63*, 323–330. [CrossRef]

19. Zhang, Y.; Guilbert, L.; Bourson, P.; Polgár, K.; Fontana, M.D. Characterization of short-range heterogeneities in sub-congruent lithium niobate by micro-Raman spectroscopy. *J. Phys. Condens. Matter* **2006**, *18*, 957–963. [CrossRef]
20. Hatano, H.; Liu, Y.; Kitamura, K. Growth and Photorefractive Properties of Stoichiometric $LiNbO_3$ and $LiTaO_3$. In *Photorefractive Materials and Their Applications 2*, 1st ed.; Günter, P., Huignard, J.P., Eds.; Springer Series in Optical Sciences: New York, NY, USA, 2007; pp. 127–164.
21. Kong, L.B.; Chang, T.S.; Ma, J.; Boey, F. Progress in synthesis of ferroelectric ceramic materials via high-energy mechanochemical technique. *Prog. Mater. Sci.* **2008**, *53*, 207–322. [CrossRef]
22. Suryanarayana, C. Mechanical alloying and milling. *Prog. Mater. Sci.* **2001**, *46*, 1–184. [CrossRef]
23. Crystallographic Open Database, Information for card entry 2101175. Available online: http://www.crystallography.net/cod/2101175.html (accessed on 29 January 2019).
24. Degen, T.; Sadki, M.; Bron, E.; König, U.; Nènert, W. The HighScore suite. *Powder Diffr.* **2014**, *29*, S13–S18. [CrossRef]
25. FIZ Karlsruhe ICSD, ICDS- Inorganic Crystal Structure Database. Available online: www2.fiz-karlsruhe.de/icsd_home.html (accessed on 29 January 2019).
26. Porto, S.P.S.; Krishnan, R.S. Raman Effect of Corundum. *J. Chem. Phys.* **1967**, *47*, 1009–1011. [CrossRef]
27. Kubelka, P. New Contributions to the Optics of Intensely Light-Scattering Materials. Part I. *J. Opt. Soc. Am.* **1948**, *38*, 448–457. [CrossRef]
28. Kubelka, P. New Contributions to the Optics of Intensely Light-Scattering Materials. Part II: Nonhomogeneous Layers. *J. Opt. Soc. Am.* **1954**, *44*, 330–335. [CrossRef]
29. Torrent, J.; Barrón, V. Diffuse Reflectance Spectroscopy. In *Methods of Soil Analysis Part 5—Mineralogical Methods*, 1st ed.; Ulery, A.L., Drees, R., Eds.; Soil Science Society of America: Wisconsin, WI, USA, 2008; pp. 367–385.
30. The Royal Society of Chemistry, Periodic Table. Available online: http://www.rsc.org/periodic-table (accessed on 29 January 2019).
31. Baird, D.C. *Experimentation: An Introduction to Measurement Theory and Experiemtn Design*, 3rd ed.; Prentice-Hall: Englewood Cliffs, NJ, USA, 1995; pp. 129–133.
32. Kalinnikov, V.T.; Gromov, O.G.; Kunshina, G.B.; Kuz'min, A.P.; Lokshin, E.P.; Ivanenko, V.I. Preparation of $LiTaO_3$, $LiNbO_3$, and $NaNbO_3$ from Peroxide Solutions. *Inorg. Mater.* **2004**, *40*, 411–414. [CrossRef]
33. Liu, M.; Xue, D.; Luo, C. Wet chemical synthesis of pure $LiNbO_3$ powders from simple niobium oxide Nb_2O_5. *J. Alloys Compd.* **2006**, *426*, 118–122. [CrossRef]
34. Scott, B.A.; Burns, G. Determination of Stoichiometry Variations in $LiNbO_3$ and $LiTaO_3$ by Raman Powder Spectroscopy. *J. Am. Ceram. Soc.* **1972**, *55*, 225–230. [CrossRef]
35. Liu, M.; Xue, D. An efficient approach for the direct synthesis of lithium niobate powders. *Solid State Ionics* **2006**, *177*, 275–280. [CrossRef]
36. Liu, M.; Xue, D.; Li, K. Soft-chemistry synthesis of $LiNbO_3$ crystallites. *J. Alloys Compd.* **2008**, *449*, 28–31. [CrossRef]
37. Nyman, M.; Anderson, T.M.; Provencio, P.P. Comparison of Aqueous and Non-aqueous Soft-Chemical Syntheses of Lithium Niobate and Lithium Tantalate Powders. *Cryst. Growth Des.* **2009**, *9*, 1036–1040. [CrossRef]
38. De Figueiredo, R.S.; Messaia, A.; Hernandes, A.C.; Sombra, A.S.B. Piezoelectric lithium niobate obtained by mechanical alloying. *J. Mater. Sci. Lett.* **1998**, *17*, 449–451. [CrossRef]
39. Pezzotti, G. Raman spectroscopy of piezoelectrics. *J. Appl. Phys.* **2013**, *113*, 211301. [CrossRef]
40. Repelin, Y.; Husson, E.; Bennani, F.; Proust, C. Raman spectroscopy of lithium niobate and lithium tantalite. Force field calculations. *J. Phys. Chem. Solids* **1999**, *60*, 819–825. [CrossRef]
41. Thermo Fisher Scientific, Application Note: Curve Fitting in Raman and IR Spectroscopy. Available online: https://www.thermofisher.com/search/results?query=Curve%20Fitting%20in%20Raman&focusarea=Search%20All (accessed on 29 January 2019).
42. Tuschel, D. The Effect of Microscope Objectives on the Raman Spectra of Crystals. *Spectroscopy* **2017**, *32*, 14–23.
43. Maïmounatou, B.; Mohamadou, B.; Erasmus, R. Experimental and theoretical directional dependence of optical polar phonons in the $LiNbO_3$ single crystal: New and complete assignment of the normal mode frequencies. *Phys. Status Solidi B* **2016**, *253*, 573–582. [CrossRef]

44. Yang, X.; Lang, G.; Li, B.; Wang, H. Raman Spectra and Directional Dispersion in LiNbO$_3$ and LiTaO$_3$. *Phys. Status Solidi B* **1987**, *142*, 287–300. [CrossRef]
45. Balanevskaya, A.É.; Pyatigorskaya, L.I.; Shapiro, Z.I.; Margolin, L.N.; Bovina, E.A. Determination of the composition of LiNbO$_3$ specimens by Raman spectroscopy. *J. Appl. Spectrosc.* **1983**, *38*, 491–493. [CrossRef]
46. Kovács, L.; Kocsor, L.; Szaller, Z.; Hajdara, I.; Dravecz, G.; Lengyel, K.; Corradi, G. Lattice Site of Rare-Earth Ions in Stoichiometric Lithium Niobate Probed by OH$^-$ Vibrational Spectroscopy. *Crystals* **2017**, *7*, 230. [CrossRef]
47. Redfield, D.; Burke, W.J. Optical absorption edge of LiNbO$_3$. *J. Appl. Phys.* **1974**, *45*, 4566–4571. [CrossRef]
48. Kovács, L.; Ruschhaupt, G.; Polgár, K.; Corradi, G.; Wöhlecke, M. Composition dependence of the ultraviolet absorption edge in lithium niobate. *Appl. Phys. Lett.* **1997**, *70*, 2801–2803. [CrossRef]
49. Thierfelder, C.; Sanna, S.; Schindlmayr, A.; Schmidt, W.G. Do we know the band gap of lithium niobate? *Phys. Satus Solidi C* **2010**, *7*, 362–365. [CrossRef]
50. Devonshire, A.F. Theory of ferroelectrics. *Adv. Phys.* **1954**, *3*, 85–130. [CrossRef]
51. O'Bryan, H.M.; Gallagher, P.K.; Brandle, C.D. Congruent Composition and Li-Rich Phase Boundary of LiNbO$_3$. *J. Am. Ceram. Soc.* **1985**, *68*, 493–496. [CrossRef]
52. Bordui, P.F.; Norwood, R.G.; Jundt, D.H.; Fejer, M.M. Preparation and characterization of off-congruent lithium niobate crystals. *J. Appl. Phys.* **1992**, *71*, 875–879. [CrossRef]
53. Kittel, C. *Introduction to Solid State Physics*, 7th ed.; John Wiley & Sons: New York, NY, USA, 1996; pp. 99–130.
54. Quin, W.; Nagase, T.; Umakoshi, Y.; Szpunar, J.A. Relationship between microstrain and lattice parameter change in nanocrystalline materials. *Philos. Mag. Lett.* **2008**, *88*, 169–179. [CrossRef]
55. Iyi, N.; Kitamura, K.; Izumi, F.; Yamamoto, J.K.; Hayashi, T.; Asano, H.; Kimura, S. Comparative study of defect structures in lithium niobate with different compositions. *J. Sol. State Chem.* **1992**, *101*, 340–352. [CrossRef]

© 2019 by the authors. Licensee MDPI, Basel, Switzerland. This article is an open access article distributed under the terms and conditions of the Creative Commons Attribution (CC BY) license (http://creativecommons.org/licenses/by/4.0/).

Article

Mechanochemical Reactions of Lithium Niobate Induced by High-Energy Ball-Milling

Laura Kocsor [1,2], László Péter [1], Gábor Corradi [1], Zsolt Kis [1], Jenő Gubicza [3] and László Kovács [1,*]

1. Wigner Research Centre for Physics, Hungarian Academy of Sciences, Konkoly-Thege Miklós út 29-33, 1121 Budapest, Hungary
2. Hevesy György PhD School of Chemistry, Eötvös Loránd University, Pázmány Péter sétány 1/A, 1117 Budapest, Hungary
3. Department of Materials Physics, Eötvös Loránd University, Pázmány Péter sétány 1/A, 1117 Budapest, Hungary
* Correspondence: kovacs.laszlo@wigner.mta.hu; Tel.: +36-1-392-2588

Received: 24 May 2019; Accepted: 26 June 2019; Published: 28 June 2019

Abstract: Lithium niobate ($LiNbO_3$, LN) nanocrystals were prepared by ball-milling of the crucible residue of a Czochralski grown congruent single crystal, using a Spex 8000 Mixer Mill with different types of vials (stainless steel, alumina, tungsten carbide) and various milling parameters. Dynamic light scattering and powder X-ray diffraction were used to determine the achieved particle and grain sizes, respectively. Possible contamination from the vials was checked by energy-dispersive X-ray spectroscopy measurements. Milling resulted in sample darkening due to mechanochemical reduction of Nb (V) via polaron and bipolaron formation, oxygen release and Li_2O segregation, while subsequent oxidizing heat-treatments recovered the white color with the evaporation of Li_2O and crystallization of a $LiNb_3O_8$ phase instead. The phase transformations occurring during both the grinding and the post-grinding heat treatments were studied by Raman spectroscopy, X-ray diffraction and optical reflection measurement, while the Li_2O content of the as-ground samples was quantitatively measured by coulometric titration.

Keywords: lithium niobate; high-energy ball-milling; nanocrystals; mechanochemical reaction

1. Introduction

Lithium niobate (LN, $LiNbO_3$) crystals have countless non-linear optical and acoustic applications due to their versatile optical and ferroelectric properties. In recent decades nanocrystalline materials have attracted considerable interest as their properties greatly differ from their coarse-grained counterparts. The reduction of the grain size may cause, e.g., increased mechanical strength, higher specific heat and larger electrical resistivity [1]. Nonlinear optical nanocrystals may have many applications such as building blocks of coherent subwavelength light sources [2–6] and nonresonant markers in second harmonic generation microscopy [4,7,8]. Rare-earth-doped LN nanocrystals could play an important role in coherent quantum optical experiments, e.g., as single photon sources, providing sharply defined wavelengths [9,10].

Many wet chemical synthesis methods are known for the preparation of nanocrystalline materials. In these methods, nanocrystals are assembled from single atoms or molecules (bottom-up methods). The other way to prepare dispersed nanograined material is to reduce the particle size by mechanical grinding (top-down method). High-energy ball-milling is one of these methods. This is a simple, general and easy-handling technique for nanocrystal preparation, while another advantage of this method is the possibility to produce large quantities of nanopowders. However, it is known that mechanical grinding may also induce phase transformations and chemical reactions beside particle and grain size reduction [11–14]. An example for a phase transformation without composition change was the occurrence of various phases in ball-milled TiO_2 (rutile, anatase and srilankite) whereas the starting material contained a single phase only [11,15]. The mechanochemical transformation can also be used for the synthesis of nanocrystals, as it was demonstrated by grinding a $(1-x)Li_2O: xB_2O_3$ mixture resulting in the formation of $Li_2B_4O_7$ [15].

Ball-milling was proved to be a successful method also for the preparation of lithium niobate. De Figueiredo et al. reported that ball-milling of Li_2CO_3 and Nb_2O_5 resulted in the formation of $LiNbO_3$, but non-reacted starter materials also remained in the milled powder [16]. Pure $LiNbO_3$ can still be prepared by mechanochemical methods if the milling process is followed by calcination at high temperature [17–19].

High-energy ball-milling was used in several works to produce nano-LN. Spex Mixer Mill 8000 is one of the most commonly used ball mills for this purpose. In general, dry grinding with one ball with a diameter of around 1 cm was used in these processes with grinding times as long as 5–20 h [13,15,20–22] and even longer than 100 h [23]. At the beginning, the grain size rapidly decreased with increasing grinding time as determined by X-ray diffraction measurements. Five and 20 h of grinding resulted in grain sizes of about 60 and 20 nm, respectively, but longer ball-milling did not decrease the grain size considerably. It should be noted that the grain size deduced from X-ray diffraction data is informative for the coherently scattering domains but does not characterize the size of the particles obtained.

Pooley and Chadwick reported that the sample ball-milled for five hours in a stainless steel vial contained a significant amount of amorphous material [13] and traces of iron [20]. Other experiments also proved that contamination from the vial's material may appear: a sample ball-milled for 16 h in an alumina vial contained 5% of alumina [22]. In these reports, however, there is no further information about whether either a phase transformation or any mechanochemical reaction occurred.

The aim of the present work is a systematic characterization of LN nanocrystals prepared by high-energy ball-milling. For the sake of simplicity, small pieces of the residue of Czochralski-grown single crystals with identical purity are used as starting material, since the milling experiments did not require large, perfect single crystalline LN. Besides the determination of both particle and grain sizes and contamination from the vial's material, the possible phase transformation and structural changes induced by the mechanochemical process are also studied using dynamic light scattering (DLS), powder X-ray diffraction (pXRD), energy-dispersive X-ray spectroscopy (EDS), Raman spectroscopy, optical diffuse reflectance measurements and coulometric acid-base titration.

2. Materials and Methods

2.1. Sample Preparation

Lithium niobate was prepared by solid-state reaction at 800 °C from Li_2CO_3 (Merck, Darmstadt, Germany, Suprapur, 99.99%) and Nb_2O_5 (Starck, Goslar, Germany, LN grade, 99.99%) raw materials by mixing them in the congruent 48.6:51.4 molar ratio. These high-purity materials contain impurities such as Fe, Al, W, relevant in our studies, in concentrations less than 1–2 ppm. Water-clear single crystals of congruent composition have been grown from platinum crucible by the Czochralski method, leaving the crucible residue also congruent and of single phase. $LiNbO_3$ nanocrystals were prepared from such polycrystalline crucible residue slowly cooled down to room temperature. Pieces of sizes of several millimeters were selected and ball-milled, using a Spex 8000 Mixer Mill (Metuchen, NJ, USA). Dry ball-milling was carried out using different types of vials and balls (SS: stainless steel, ALO: alumina, WC: tungsten carbide). The grinding time was 5 or 20 h with interruptions of 30 min after every hour of grinding to avoid overheating (the ball mill was not equipped with a temperature sensor). Further milling parameters are given in Table 1. During the milling process and thereafter the samples were kept in air (no protecting atmosphere was used).

After the milling process, some samples showed greyish coloration usually observed for LN single crystals after reduction (i.e., decomposition with oxygen release and a concomitant formation of Nb (IV) in the lattice). This indicates that the samples underwent a change concerning the oxidation state of niobium during the grinding process. For this reason, annealing was applied for restoring or modifying the oxidation state of niobium in the ground samples. Annealing treatments in either oxidative or non-oxidative atmospheres were performed for 3 h at 800 °C in either air or in vacuum of about 10^{-4} mbar, respectively. Samples were put into the furnace in a form so that any possible reaction between the material and its containment could be avoided. Powder samples were held in a platinum crucible for the oxidative treatments, while pellets were placed on a platinum plate inside the quartz tube used for evacuation in the case of the application of non-oxidative conditions. Samples subjected to one of these annealings will be called hereafter oxidized and reduced samples. Oxidized-reduced and reduced-oxidized samples underwent two subsequent heat treatment processes separated by periods long enough to let the samples cool down to room temperature. This way, a series of five differently treated samples was obtained for each vial material (see Scheme 1).

Table 1. Grinding parameters used for high-energy ball-milling.

Sample	Grinding Parameters						
	Vial	Ball (Same Material as the Vial)	Time (h)	Number of Balls	Ball-to-Powder Mass Ratio	Ball-to-Powder Volume Ratio	Sample Quantity (g)
SS-5	Stainless steel	11 mm 5.5 g	5	2	3.8:1	2.2:1	2.9
ALO-5 ALO-20	Alumina	12.5 mm 4.2 g	5 20	2	3.8:1	4.4:1	2.2
WC-5	Tungsten carbide	11 mm 10.7 g	5	2	3.8:1	1.2:1	5.65

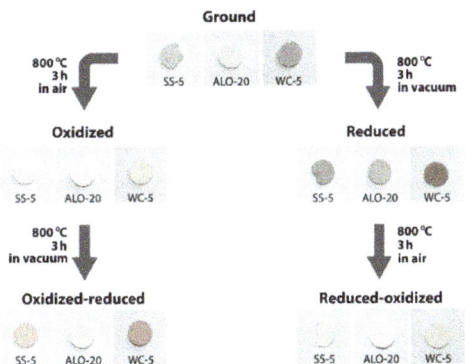

Scheme 1. The series of samples ground in different vials (see Table 1) with subsequent heat treatments. Pellets were pressed for optical reflectance measurements.

2.2. Sample Characterization

The phase analysis of the samples was carried out by X-ray diffraction (XRD) using a Rigaku (Tokyo, Japan) Smartlab X-ray diffractometer with CuKα radiation (wavelength: λ = 0.15418 nm). The XRD patterns were measured in the Bragg-Brentano diffraction geometry. The crystalline phases were identified from the peak positions and intensities using an ICDD PDF-2 database. The grain sizes were determined by using the Williamson-Hall method.

The Raman spectra of ground and heat-treated samples were collected at room temperature using a Renishaw (Wotton-under-edge, UK) inVia Raman Microscope in backscattering geometry. A 633 nm laser beam was used as excitation source using a 50× lens. The excitation spot size was 2 μm at the sample surface. The Raman data were recorded in the range of 20–460 cm^{-1} with a low-wavenumber filter and in the range of 100–1200 cm^{-1} with a notch filter.

Dynamic light scattering measurements in the range of 0.1–10000 nm (Malvern Zetasizer Nano S, Worcestershire, UK) were performed to determine the particle size distribution. The ground samples were suspended in water assuming the nanoparticles to be perfect spheres leaving the viscosity of water unchanged. The refractive index of the particles was taken as ~2.2 corresponding to that of bulk $LiNbO_3$ in the given wavelength region.

Since the LN samples were discolored after the grinding procedure, possible contamination from the vials was suspected. This was checked by EDS measurements in a Zeiss Leo scanning electron microscope (Jena, Germany). Depending on the vial used, the presence of Fe, Al and W was scrutinized for the stainless steel, alumina and tungsten carbide vials, respectively. The sensitivity of the EDS method for the above listed elements is 0.1 at%, while the relative error of the EDS measurement in the $x < 0.5$ at% concentration range can be as high as 20%.

To characterize the optical reflectance properties of the samples, an Avantes (Apeldorn, The Netherlands) HS-1024X122 TEC UV-VIS fiber-optic modular spectrophotometer was used. The measurements were carried out on pellets pressed from the ground and heat-treated samples. The samples were illuminated in the 190–1100 nm wavelength range using an Avantes AvaLight-D(H)-S deuterium-halogen light source. To determine the reflectance spectra, we used an Avantes WS-2 diffuse white tile as a reference. The spectral data recorded were used without any further processing.

The secondary phase occurring during the grinding process (Li_2O or any other compound it may be transformed to) was quantitatively measured by semi-micro coulometric acid-base titration with current control. For this procedure, portions of the powder samples were weighted in a microbalance with a precision of 2 µg and suspended in a few milliliters of 0.3 mol/dm^3 Na_2SO_4 solution in the anode compartment of a diaphragm-divided two-chamber electrochemical cell. The pH in the anode compartment was measured with a combined glass pH electrode and a Consort 860 pH tester. A platinum anode was used for acid generation by the following electrode reaction: $2H_2O = O_2 + 4H^+ + 4e-$. The Na_2SO_4 solution was boiled prior to the measurement in order to remove any dissolved CO_2 that may interfere with the measurements, then cooled to room temperature for the titration. The solution in the anode compartment was stirred with a magnetic stirrer bar. Blank measurements and known quantities of both Na_2HPO_4 and NaOH were used to verify that no solution intermixing took place between the anode and cathode chambers.

3. Results and Discussion

3.1. Result of the Ball Milling: Size Parameters

Figure 1 shows the as-received particle diameter distributions of the ground samples determined by the DLS method. All of them have a broad distribution in the range of a few hundred nanometers while the smallest particle diameter can be achieved by the ball-milling in stainless steel vial. Ball-milling in alumina or tungsten carbide vial resulted in similar particle diameter distributions which did not change considerably if a milling time longer than 5 h was chosen.

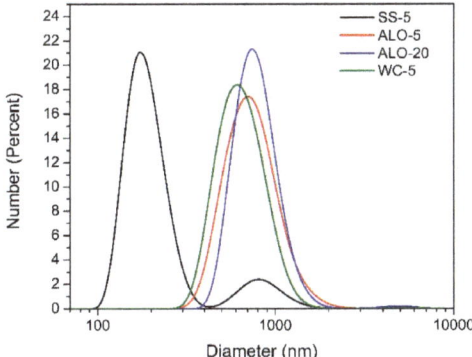

Figure 1. Particle diameter distributions of the ground samples in different vials determined by the dynamic light scattering (DLS) method.

Since the samples darkened as a result of the grinding process, it was checked whether the greyish color was related to sample contamination originating from the ball or the vial. EDS results showed that no characteristic impurity element at $x > 0.1$ at% concentration was present in the LN powders ground in either the tungsten carbide or the stainless steel vials, while aluminum could be detected when alumina vial was used in the milling process. This also means that the large fraction corresponding to smaller particle sizes observed in sample SS-5 milled in the stainless steel vial (see Figure 1) cannot be ascribed to the presence of an impurity but has to be considered as LN, too.

Neither the occurrence of alumina contamination nor the absence of any stainless steel residue can be explained on the basis of the hardness properties of the materials involved in the ball milling process. Taking into account the order of the Vickers hardness values of the relevant materials (stainless steel: 200–240, LN: ~630 [24], tungsten carbide: 1200–1700, alumina: 1400–1900), one might expect that stainless steel would cause the largest impurity level. However, alumina was the only vial material resulting in sample contamination, although it was the hardest material used. Al contamination was also confirmed by the XRD measurements as indicated by the presence of α-Al_2O_3 in the XRD pattern. This phenomenon was also proved by Heitjans and co-workers [22]. Although the alumina impurity level was sufficiently high for both XRD and EDS detection (at least 2 vol% which was the detection limit of the XRD method), it did not appear as a discernible fragment in the size distribution curves of the samples ALO-5 and ALO-20 (see Figure 1).

Although the alumina contamination cannot be explained on the basis of the hardness of the unmodified materials, it is understood by taking into account the mechanochemical reactions taking place in the system. As it will be shown in the forthcoming sections, lithium oxide is released from LN during the milling process that can react with the amphoteric alumina, resulting in the modification of its structure and facilitating its degradation. Various mixed oxide compounds of aluminum and lithium are known (such as $LiAlO_2$, $LiAl_5O_8$, Li_3AlO_4; see Ref. [25]). In particular, the synthesis of $LiAlO_2$ is well described by using either sol-based [26] or solid phase reaction-based [27] routes. Although the quantitative yield of the solid-phase reaction is given for temperatures higher than 370 °C, a surface-limited reaction can be assumed for the milling conditions applied in the present work. The hardness data of $LiAlO_2$ is not known to the authors, but the Vickers hardness of other alkali aluminates (Na or K) is reported to be quite small (< 100, see [28]). Hence, the sites where the released Li_2O reacts with the Al_2O_3 ball/vials can be assumed to serve as degradation initiation spots during the milling process.

While the particle size of a few hundred nm observed by light scattering did not diminish further for milling times longer than 5 h, a Williamson-Hall type evaluation of the X-ray diffractograms resulted in mean grain sizes decreasing from at about 63 nm to 37 nm for milling times increasing from 5 to 20 h (see Table 2). This reflects the fact that the diffraction-based size analysis provides the diameter of the coherently scattering grains which differs from the much larger particle diameter visible by light scattering or imaging methods. This is direct evidence that the particles formed during high-energy milling consist of a multitude of small grains. The grain diameters found in the present work are in good agreement with those reported before [13,15,20–23] for ball-milled LN crystals.

Table 2. Particle and grain sizes of samples ground in different vials.

Sample	Resulting Particle Diameter (nm) DLS	Resulting Grain Diameter (nm) XRD
SS-5	190, (800)	55 ± 18
ALO-5	700	63 ± 21
ALO-20	700	37 ± 2
WC-5	500	51 ± 6

3.2. Structure of the Samples: XRD, Raman and Optical Reflectance Measurements

Figure 2 shows the XRD patterns of the ground materials compared with those oxidized and/or reduced at 800 °C for 3 h. The broad peaks of the XRD patterns of the as-ground samples indicate small grain sizes as determined numerically using the Williamson-Hall method (see Table 2). Heat-treatment processes resulted in narrower diffraction lines due to increased grain sizes. The reflections of a $LiNb_3O_8$ (lithium triniobate) phase appeared in the diffraction patterns of all annealed samples (best seen for oxidized samples, especially for those ground in the stainless steel vial, see Figure 2a). The formation of the $LiNb_3O_8$ phase taking place as a result of the combined ball-milling and annealing procedure can be described as

$$3\ LiNbO_3 = LiNb_3O_8 + Li_2O, \qquad (1)$$

where lithium oxide is a volatile byproduct.

The fact that the lithium triniobate can only be identified in the heat-treated samples indicates that the structural rearrangement of the residual Nb containing oxides is not completed in the as-ground samples. However, the annealing process provides the activation energy required for crystallization of the new phase with sufficiently large crystallites to yield strong enough reflections in the diffraction patterns.

In the XRD pattern of the sample ball-milled in alumina vial the reflections of α-Al_2O_3 can be clearly identified due to abrasion of the vial and balls during the milling process already before the heat-treatments (see Figure 2b). The fact that alumina was present in a crystalline form already in the as-ground samples indicates that the milling destroyed the ball/vial material and the majority of this impurity does not arise as a result of the side reaction that weakened the alumina structure. The lack of further crystalline compounds of aluminum in the annealed samples shows that the amount of possibly reacted alumina was insignificant as compared to alumina that entered the ground mixture by the mechanical effect of milling.

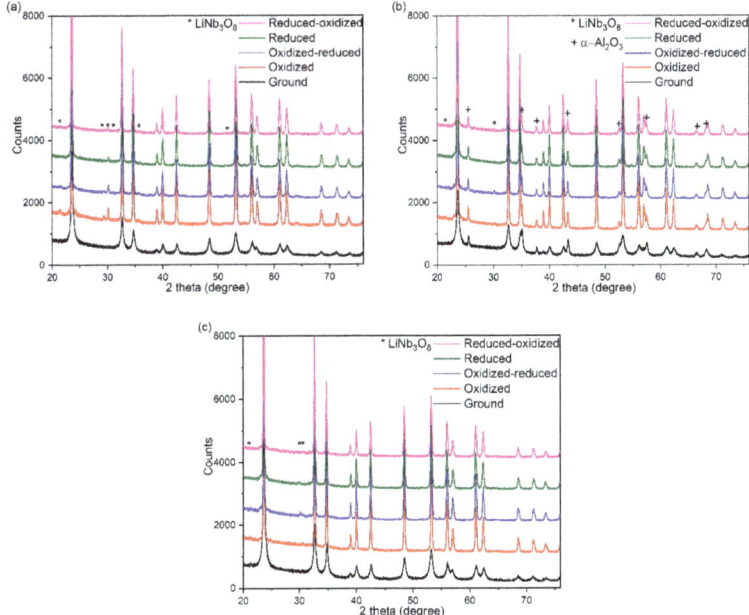

Figure 2. Diffraction patterns of Lithium niobate (LN) ground in different vials, stainless steel (**a**), alumina (**b**), and tungsten carbide (**c**). The unmarked peaks are the reflections of $LiNbO_3$.

Figure 3 shows the Raman spectra of the as-ground and heat-treated samples ball-milled in stainless steel and tungsten carbide vials. The Raman intensities of the as-ground samples are weak, the bands are broad, not showing all characteristic features of $LiNbO_3$ crystals. The heat-treatment process resulted in line narrowing and increased intensity of the bands corresponding to the pure $LiNbO_3$ phase. In addition, in the oxidized samples some weak bands appeared at 59, 79 and 96 cm^{-1} corresponding to the $LiNb_3O_8$ phase (Figure 3a) [29]. This confirms the XRD results, where the presence of the $LiNb_3O_8$ phase predicted by Equation (1) was best seen for oxidized samples, especially for those ground in stainless steel vial.

Equation (1) suggests that at least one new component without any niobium content has to appear during the milling process. Lithium oxide, Li_2O, may be present as the primary byproduct and can be transformed in air to another lithium compound ($LiOH·xH_2O$, $Li_2CO_3·xH_2O$) by water and/or CO_2 uptake. Indeed, the water suspensions of all ball-milled LN powders were found to be alkaline, regardless of the chemical state of the Li-rich segregate, which is an unambiguous confirmation of the decomposition of LN via Li_2O separation during the milling process. The as-ground LN particles were structurally disordered in the decomposed region but recrystallized upon annealing, hence both the Raman and XRD lines of $LiNb_3O_8$ could manifest themselves. The CO_2 uptake of Li_2O produced during ball-milling can also be observed, viz. in the Raman spectrum of the LN powder ground in tungsten carbide vial shown in Figure 3b. The bands at about 190 and 1090 cm^{-1} present in the as-ground samples are attributed to Li_2CO_3 generated from Li_2O (Figure 3b) [30]. Heat-treatments at 800 °C either in air or in vacuum resulted in the loss of CO_2 evidenced by the disappearance of those bands from the Raman spectra. The presence of α-Al_2O_3 contamination in the powder ball-milled in alumina vial was observed in the XRD diffractogram; however, it could not be detected by Raman spectroscopy as the corresponding bands at about 383 and 420 cm^{-1} overlap with the larger bands of LN [31].

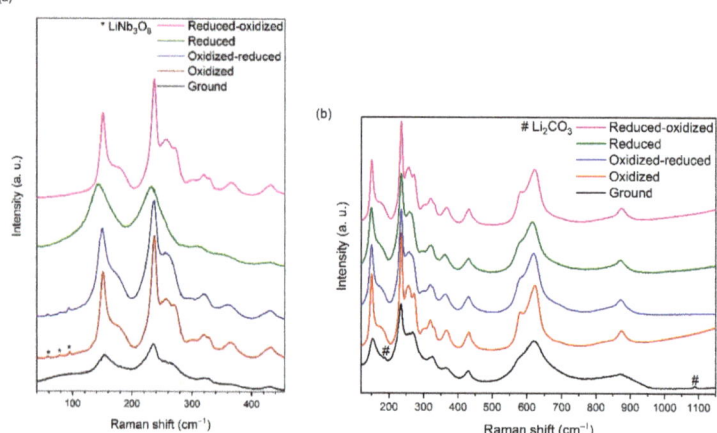

Figure 3. Raman spectra of ground and heat-treated samples ball-milled for five hours in stainless steel vial, shown in the range of 40–460 cm^{-1} (**a**), and in tungsten carbide vial, shown between 125–1150 cm^{-1} (**b**).

Equation (1) does not account for the redox processes indicated by color changes seen during ball-milling and annealing treatments. The colors of samples ground in alumina, stainless steel and tungsten carbide vials varied from light gray to dark gray (see Scheme 1), evidenced by their optical reflection spectra (Figure 4a)—the darker the sample, the lower its reflection in the whole spectral range. As mentioned above, the samples underwent a change concerning the oxidation state of niobium during the grinding process. This partial reduction could be compensated by oxidizing the sample applying a heat-treatment in air at 800 °C. The oxidative annealing resulted in white color for the powder ground in stainless steel vial (Scheme 1). Upon subsequent reduction, the sample became brownish, while the pellet pressed from the as-ground powder became gray when reduced directly. The oxidation process resulted in a white color even in the case of the previously reduced sample. As an example, the optical reflection spectra of the as-ground and annealed samples ball-milled in stainless steel vial are shown in Figure 4b. Similar effects were observed for powders ground in the other two vials: the change of color was less evidenced for alumina vial but was stronger for tungsten carbide vial.

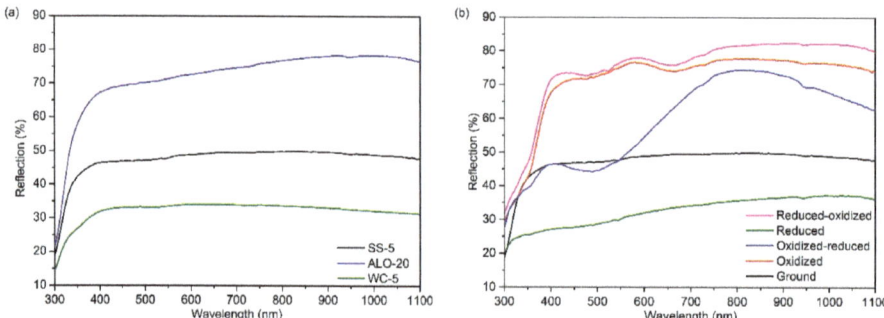

Figure 4. Optical reflection spectra of samples ball-milled in different vials (**a**) and of ground and heat-treated samples ball-milled in stainless steel vial (SS-5) (**b**).

In Figure 4a no distinctive feature characteristic for Fe^{2+} having an absorption band near 500 nm can be discerned for any of the as-ground samples. The comparison of reduced samples also shows very small differences in this respect. Instead, differences in amplitude of the whole spectrum dominate. Still some change of coloration induced by the redox treatments may be related to the iron contamination at least partly coming from the starting material. For the preparation of $LiNbO_3$ high- purity raw materials with less than 2 ppm, Fe was used. Since the effective distribution coefficient of Fe between the molten and solid lithium niobate of congruent composition is around 1, it is not expected to be enriched in the crucible residue during the growth process. On the other hand, even 10 ppm iron does not induce a sizable increase of the optical absorption. According to Phillips et al. [32] the difference in the absorption coefficient at about 500 nm between oxidized and reduced LN containing about 0.5 mol% Fe is less than 4 cm^{-1}, i.e. less than 0.001 cm^{-1}/ppm Fe, which cannot cause dominant coloration changes in our case.

3.3. Discussion of the Mechanochemical Reaction Including Redox Processes

Congruent $LiNbO_3$ crystals are strongly Li deficient and can be described by the $Li_{1-5x}Nb_{1+x}O_3$ formula, where $x \approx 0.01$. The excess Nb ions occupy Li sites and are charge compensated by Li vacancies. The antisite Nb_{Li} ions may trap electrons, forming small polarons. Moreover, Nb_{Li}—Nb_{Nb} pairs, consisting of an antisite and its regular nearest-neighbor along the ferroelectric c axis, are capable to form stable bipolarons (for a review see [33]). The strongly localized polaron/bipolaron models can evidently be applied for the redox processes in LN nanocrystals.

Already before the appearance of the $LiNb_3O_8$ phase the LN particles underwent partial reduction as a result of ball-milling. During reduction, oxygen gas and lithium oxide are formed, the latter leaving the sample only upon annealing treatments. This leads to the appearance of various polarons with elementary cell loss at the surface [33,34]

$$2LiNbO_3 \rightarrow Nb_{Nb}^{4+} + Nb_{Li}^{4+} + 3O_O^{2-} + 2e^- + Li_2O\uparrow + O_2\uparrow \qquad (2)$$

where the $Nb_{Nb}^{4+} + Nb_{Li}^{4+}$ pair makes a bipolaron. The remaining electrons may also form either a further bipolaron or two Nb_{Li}^{4+} polarons on pre-existing Nb_{Li} antisite defects in the congruent bulk. Their broad absorption bands near 500 nm and 760 nm and the disordered structure result in uniform gray color as observed for the as-ground samples (see Scheme 1 and reflection spectra in Figure 4a). In this stage the surface is disordered and consists of strongly subcongruent lithium niobate, while Li_2O forms a different phase.

The structure of the particles consisting of a large number of grains can be understood to result from an interplay of disturbed ferroelectric surface fields. The grains may be assumed to be monodomain regions of LN pairwise attracted by the strong electric fields acting on most blank surfaces of this ferroelectric. Structural damage does not allow exact fitting of the attached surfaces, resulting also in crystallographic misorientation of the grains in touch. During ball milling, the particles constantly break up and recoalesce in different arrangements, while reduction may only proceed on surfaces where oxygen evaporation is possible for a sufficiently long period. Li_2O segregation on such exposed grain surfaces may finally shield the electric fields and hamper further attachments with neighboring grains. During prolonged grinding, this may result in a structure where a large part of the Li_2O phase is on the external surface of the particles.

Oxidation of the samples leads to the evaporation of the segregated Li_2O phase and in parallel the disappearance of all polarons. The latter recovers the white color by reverting the reduction described by Equation (2) and promotes the formation of the $LiNb_3O_8$ phase according to Equation (1) whereby a further LN formula unit is used up:

$$LiNbO_3 + Nb_{Nb}^{4+} + Nb_{Li}^{4+} + 3O_O^{2-} + 2e^- + O_2 \rightarrow LiNb_3O_8 \qquad (3)$$

as observed by XRD, Raman and optical reflection measurements prominently for the oxidized samples. The LiNb$_3$O$_8$ phase may form an epitaxial layer on the LiNbO$_3$ surface as described by Semiletov et al. [35].

In all other preparation stages various mixtures of bipolarons and polarons are present mainly absorbing in the blue-green and red range, respectively (Figure 4b). While the as-ground state has a balanced mixture, its direct reduction leads to a larger bipolaron portion (less reflection in the blue-green region, see the curve with lowest reflection in Figure 4b. The same reduction, if preceded by oxidation, reproduces only bipolarons but very few Nb$_{Li}^{4+}$ polarons (high reflection only in the red region, see blue curve). Some additional structure observed near 350 nm and 670 nm in samples having an oxidizing step in their history might be attributed to absorption related to the LiNb$_3$O$_8$ phase. As shown by Sugak et al. [36] the coloration is formed near the crystal surface and its distribution depends on annealing temperature. Annealing is assumed to attack the exposed surface of the particles without essentially changing their deeper structure. It should be noted that the large formation enthalpy of oxygen vacancies in LN compared to that of similar defects of the cation sublattice prevents the diffusion of oxygen within the bulk, while diffusing cations may easily occupy the empty Li sites amply available in congruent LN (see [33] and references therein).

Reaction (2) is an equivalent version of Equation (2) in Sugak et al.'s paper, separately showing near-surface formation of polarons by Nb displacement to a Li site on the one hand, and electrons available for diffusion to more deeply situated antisites causing similar coloration on the other hand. This distinction, together with the overlooked fact that elementary cells are lost upon reduction, resolves the problems of Sugak et al. about unrealistic properties of coloration allegedly following from their Equation (2). In contrast to the opinion of Sugak et al., reduction-oxidation cycles are not completely reversible processes due to possible Li oxide loss especially at higher temperatures and in closely stoichiometric LiNbO$_3$, the latter being also much more resistant to reduction. Thermal reduction was shown to increase off-stoichiometry which, in turn, leads to larger density [37,38], quantitatively supporting Equation (2). All this gives full support to the cationic model of coloration excluding any diffusion of oxygen in the bulk. A further argument for reaction (2) specifically in our case is the expected higher density of Li-poor grain kernels, taken into account that they are produced by mechanical pressing exerted by the vials.

3.4. Quantitative Determination of the Degree of Decomposition During Ball-Milling

The quantity of lithium oxide segregated at the particle surfaces during ball-milling was determined by coulometric titration in the as-ground samples. No similar measurements were attempted for the annealed samples since the Li$_2$O has a fairly large volatility at the annealing temperature. All titration curves exhibited a single neutralization step as the acid was produced in-situ by the current passing through the cell. This indicates that the primary decomposition product was Li$_2$O and no significant amount of Li$_2$CO$_3$ was present, despite long storing times of several weeks in air elapsed after grinding prior to the titration procedure. The presence of carbonate should have led to a two-stage neutralization process, first leading to bicarbonate formation, but this was never observed. Although the decomposition product detected by the Raman measurement was lithium carbonate, this is no counterargument, as the Raman intensity of the various lithium-containing compounds may be very different and hence, the sensitivity of the Raman measurement may not be comparable for the various possible phases. The weight of dissolved Li$_2$O (m_{OX}) was calculated with the following equation:

$$m_{OX} = MQ/2F \qquad (4)$$

where M is the molar weight of Li$_2$O, Q is the charge passed until the equivalence point, F is the Faraday-constant (96485 C/mol), while the number in the denominator indicates that the hydrolysis of 1 mol of Li$_2$O results in 2 mols of hydroxide ions. Four measurements were performed for each batch ground in different vials. The measured Li$_2$O mass ratios w_{ox}, expressed as weight percent of the as-ground powder are given in Table 3. The uncertainty of the measurements is given as the standard deviation of the consecutive titration results.

Table 3. Measured and estimated parameters of the nanocrystalline LiNbO$_3$ samples ground in different vials. The weight of the Li$_2$O segregate was measured by titration before annealing, while the LiNb$_3$O$_8$ shell crystallized only upon annealing.

Sample	100w_{ox} Li$_2$O in As-Ground Samples (Weight%)	R Mean Particle Radius, DLS (nm) **	r Mean Grain Radius, XRD (nm) **	d_{LTN} LiNb$_3$O$_8$ Shell Thickness (nm)	d_{OX} Li$_2$O Shell Thickness (nm)
ALO-20 *	0.97 ± 0.05	350	18.5	14.6	2.6
WC-5	1.05 ± 0.10	250	25.5	11.3	2.0
SS-5	1.52 ± 0.21	95 ***	27.5	6.2	1.1

* Alumina contamination neglected. ** Taken for convenience from Table 2. *** Smaller peak in the size distribution neglected.

The total surface of particles in the sample is proportional to $1/R$, where R is the average particle radius as measured by DLS. The values of w_{OX} in Table 3 indeed increase monotonously with $1/R$, though a fully quantitative trend cannot be established. No similar trend related to the inverse grain radius $1/r$ can be seen in the given range of r values obtained by XRD. These observations can be understood if segregation mainly occurs on the outer particle surfaces where both Li$_2$O and O$_2$ may freely leave, giving rise to a niobium-rich layer. However, it is also possible that part of the newly created surfaces, together with part of the Li$_2$O formed, gets buried during later stages of milling and cannot be dissolved for titration.

The quantitative determination of the lithium oxide loss enables us to estimate the thickness of the lithium triniobate layer in the oxidized samples. We use a simplified model of compact, uniform, spherical LN particles of unmodified composition covered by a uniformly thick LiNb$_3$O$_8$ phase. The shell thickness is calculated with the assumption that only the Li$_2$O equivalent to this outer shell could be dissolved and analyzed by titration.

From the reaction indicated in Equation (1) it follows that

$$n_{OX} = n_{LTN} \tag{5}$$

where n stands for the molar quantity of the relevant materials, while the indices OX and LTN refer to the lithium oxide segregate and the lithium triniobate shell of the particles, respectively. For the weight of the particle shell we obtain

$$m_{OX} = m_{LTN} M_{OX}/M_{LTN} \tag{6}$$

where M is the molar weight. The weight ratio of the lithium oxide in the ground material, w_{OX}, is as follows:

$$w_{OX} = \frac{m_{LTN} \frac{M_{OX}}{M_{LTN}}}{m_{LN} + m_{LTN}\left(1 + \frac{M_{OX}}{M_{LTN}}\right)} \tag{7}$$

The weight of each particle component can be expressed with the geometric parameter of the core-shell structure, d being the shell thickness and ρ the density:

$$m_{LN} = \frac{4}{3}\pi(R - d_{LTN})^3 \rho_{LN} \approx \frac{4}{3}\pi R^3 \rho_{LN} - 4\pi R^2 d_{LTN} \rho_{LN} \tag{8}$$

$$m_{LTN} \approx 4\pi(R - d_{LTN})^2 d\rho_{LTN} \approx 4\pi R^2 d_{LTN} \rho_{LTN} \tag{9}$$

The higher order terms with respect to d_{LTN} have been neglected since $d_{LTN} \ll R$. The weight ratio of the lithium oxide is then

$$w_{OX} = \frac{d_{LTN}\rho_{LTN}\frac{M_{OX}}{M_{LTN}}}{\frac{R}{3}\rho_{LN} + d_{LTN}\left[\rho_{LTN}\left(1 + \frac{M_{OX}}{M_{LTN}}\right) - \rho_{LN}\right]} \quad (10)$$

Again, the $d_{LTN} \ll R$ relation justifies the neglection of the second term in the denominator, leading to

$$d_{LTN} \approx \frac{R}{3}\frac{\rho_{LN}}{\rho_{LTN}}\frac{M_{LTN}}{M_{OX}}w_{OX} \quad (11)$$

By assuming that the Li$_2$O leaving the particle also forms a similar shell in the as-ground sample, for the thickness of this shell we calculate

$$d_{OX} \approx \frac{\rho_{LTN}}{\rho_{OX}}\frac{M_{OX}}{M_{LTN}}d_{LTN} \approx \frac{d_{LTN}}{5.6} \quad (12)$$

where we take $\rho_{OX} \equiv \rho_{Li_2O} \approx 2.01$ g/cm^3, $\rho_{LN} = 4.65$ g/cm^3 and $\rho_{LTN} = 4.975$ g/cm^3 [39]. The values of d_{LTN} and d_{OX} are also included in Table 3 and correspond to a layer thickness of at most a few unit cells.

The same amount of segregate (either Li$_2$O or LiNb$_3$O$_8$), if spread evenly on all grain boundaries, would result in a much thinner layer. Neither the corresponding LiNb$_3$O$_8$ layer would be seen as XRD peaks nor would the equivalent amount of Li$_2$O be readily soluble due to its hindered accessibility.

This finding gives further support to our previous assumption that the processes described by Equations (2) and (3) essentially occur on the outer surfaces. Accordingly, particle and grain size reduction proceeds as long as surfaces freshly broken up during ball milling have enough time to pile up a non-ferroelectric surface layer preventing them from stable recoalescing. Below a certain size limit, depending on the detailed properties of the milling system, this becomes impossible as recoalescence becomes too fast. The thickness of the outer segregate layer apparently has a narrow range defined by a similar requirement of sufficient atmospheric contact of the polar surface.

The proposed formation of the core-shell structure would require direct experimental evidence. However, the particle size achieved by the ball-milling process was too large for a direct transmission electron microscopic study of the particles.

Finally, we remark that the given description corresponds to the surface-screening mechanisms in ferroelectric thin films reviewed by Kalinin [40]. In particular, very similar processes seem to occur in prospective lithium-ion batteries using LiNb$_3$O$_8$ as an anode material [41].

4. Conclusions

Nano-LN was prepared by ball-milling using a Spex 8000 Mixer Mill with different milling parameters. The resulting particle size has a broad distribution in the range of a few hundred nanometers. Five and 20 h of ball-milling resulted in mean grain sizes of about 60 and 40 nm, respectively. Longer ball-millings do not decrease the particle size but only reduce the grain size. α-Al$_2$O$_3$ contamination was found for the sample ground in alumina vial due to the chemically induced abrasion of the vial and the balls during ball-milling. During the milling process the material suffers partial reduction that leads to a balanced formation of bipolarons and polarons yielding gray color together with Li$_2$O segregation on the open surfaces. Upon high-temperature oxidation, the volatile Li$_2$O phase and the polarons get eliminated and the Li deficiency is accommodated by the formation of a more stable LiNb$_3$O$_8$ shell. Darker or brownish color appearing upon high-temperature reduction is caused by the preferential formation of bipolarons. The Li$_2$O loss was observed to increase with the growing total surface of the particles. The average thickness of the non-ferroelectric surface segregate corresponds to a layer of a few unit cells forming the passivating shell of the particles. These findings provide a comprehensive explanation of the physicochemical behavior of the system during grinding and annealing in different atmospheres.

Author Contributions: Conceptualization, L.K. (László Kovács), L.P. and G.C.; methodology, L.K. (Laura Kocsor) and J.G.; formal analysis, L.K. (Laura Kocsor); investigation, L.K. (Laura Kocsor); writing—original draft preparation, L.K. (László Kovács), L.P. and G.C.; visualization, L.K. (László Kovács); supervision, L.P.; project administration, Z.K.; funding acquisition, Z.K.

Funding: This research was supported by the National Research, Development and Innovation Fund of Hungary within the Quantum Technology National Excellence Program (Project No. 2017-1.2.1-NKP-2017-00001) and the Ministry of Human Capacities of Hungary within the ELTE University Excellence program (1783-3/2018/FEKUTSRAT).

Acknowledgments: The authors are grateful to Gábor Piszter and Levente Illés for the optical diffuse reflectance and EDS measurements, respectively.

Conflicts of Interest: The authors declare no conflict of interest.

References

1. Kar, S.; Logad, S.; Choudhary, O.P.; Debnath, C.; Verma, S.; Bartwal, K.S. Preparation of Lithium Niobate Nanoparticles by High Energy Ball Milling and Their Characterization. *Univers. J. Mater. Sci.* **2013**, *1*, 18–24.
2. Bonacina, L.; Mugnier, Y.; Courvoisier, F.; Le Dantec, R.; Extermann, J.; Lambert, Y.; Boutou, V.; Galez, C.; Wolf, J.P. Polar Fe(IO$_3$)$_3$ Nanocrystals as Local Probes for Nonlinear Microscopy. *Appl. Phys. B Lasers Opt.* **2007**, *87*, 399–403. [CrossRef]
3. Zielinski, M.; Oron, D.; Chauvat, D.; Zyss, J. Second-Harmonic Generation from a Single Core/Shell Quantum Dot. *Small* **2009**, *5*, 2835–2840. [CrossRef] [PubMed]
4. Tripathi, S.; Davis, B.J.; Toussaint, K.C.; Carney, P.S. Determination of the Second-Order Nonlinear Susceptibility Elements of a Single Nanoparticle Using Coherent Optical Microscopy. *J. Phys. B At. Mol. Opt. Phys.* **2011**, *44*, 015401. [CrossRef]
5. Nakayama, Y.; Pauzauskie, P.J.; Radenovic, A.; Onorato, R.M.; Saykally, R.J.; Liphardt, J.; Yang, P. Tunable Nanowire Nonlinear Optical Probe. *Nature* **2007**, *447*, 1098–1101. [CrossRef] [PubMed]
6. Le, X.L.; Brasselet, S.; Treussart, F.; Roch, J.F.; Marquier, F.; Chauvat, D.; Perruchas, S.; Tard, C.; Gacoin, T. Balanced Homodyne Detection of Second-Harmonic Generation from Isolated Subwavelength Emitters. *Appl. Phys. Lett.* **2006**, *89*, 121118.
7. Hsieh, C.L.; Grange, R.; Pu, Y.; Psaltis, D. Three-Dimensional Harmonic Holographic Microcopy Using Nanoparticles as Probes for Cell Imaging. *Opt. Express* **2009**, *17*, 2880–2891. [CrossRef]
8. Kijatkin, C.; Eggert, J.; Bock, S.; Berben, D.; Oláh, L.; Szaller, Z.; Kis, Z.; Imlau, M. Nonlinear Diffuse Fs-Pulse Reflectometry of Harmonic Upconversion Nanoparticles. *Photonics* **2017**, *4*, 11. [CrossRef]
9. Aharonovich, I.; Castelletto, S.; Johnson, B.C.; McCallum, J.C.; Simpson, D.A.; Greentree, A.D.; Prawer, S. Chromium Single-Photon Emitters in Diamond Fabricated by Ion Implantation. *Phys. Rev. B* **2010**, *81*, 121201. [CrossRef]
10. Kolesov, R.; Xia, K.; Reuter, R.; Stöhr, R.; Zappe, A.; Meijer, J.; Hemmer, P.R.; Wrachtrup, J. Optical Detection of a Single Rare-Earth Ion in a Crystal. *Nat. Commun.* **2012**, *3*, 1027–1029. [CrossRef]
11. Begin-Colin, S.; Girot, T.; Mocellin, A.; Le Caër, G. Kinetics of Formation of Nanocrystalline TiO$_2$ II by High Energy Ball-Milling of Anatase TiO$_2$. *Nanostructured Mater.* **1999**, *12*, 195–198. [CrossRef]
12. Girot, T.; Bégin-Colin, S.; Devaux, X.; Le Caër, G.; Mocellin, A. Modeling of the Phase Transformation Induced by Ball Milling in Anatase TiO$_2$. *J. Mater. Synth. Process.* **2000**, *8*, 139–144. [CrossRef]
13. Pooley, M.J.; Chadwick, A.V. The Synthesis and Characterisation of Nanocrystalline Lithium Niobate. *Radiat. Eff. Defects Solids* **2003**, *158*, 197–201. [CrossRef]
14. Gutman, E.M. *Mechanochemistry of Materials*; Cambridge International Science Publishing: London, UK, 1998.
15. Indris, S.; Bork, D.; Heitjans, P. Nanocrystalline Oxide Ceramics Prepared by High-Energy Ball Milling. *J. Mater. Synth. Process.* **2000**, *8*, 245–250. [CrossRef]
16. De Figueiredo, R.S.; Messai, A.; Hernandes, A.C.; Sombra, A.S.B. Piezoelectric Lithium Niobate Obtained by Mechanical Alloying. *J. Mater. Sci. Lett.* **1998**, *17*, 449–451. [CrossRef]
17. Luo, J.H. Preparation of Lithium Niobate Powders by Mechanochemical Process. *Appl. Mech. Mater.* **2011**, *121–126*, 3401–3405. [CrossRef]

18. Diaz-Moreno, C.A.; Farias-Mancilla, R.; Elizalde-Galindo, J.T.; González-Hernández, J.; Hurtado-Macias, A.; Bahena, D.; José-Yacamán, M.; Ramos, M. Structural Aspects LiNbO$_3$ Nanoparticles and Their Ferromagnetic Properties. *Materials* **2014**, *7*, 7217–7225. [CrossRef]
19. Fierro-Ruiz, C.; Sánchez-Dena, O.; Cabral-Larquier, E.; Elizalde-Galindo, J.; Farías, R. Structural and Magnetic Behavior of Oxidized and Reduced Fe Doped LiNbO$_3$ Powders. *Crystals* **2018**, *8*, 108. [CrossRef]
20. Chadwick, A.V.; Pooley, M.J.; Savin, S.L.P. Lithium Ion Transport and Microstructure in Nanocrystalline Lithium Niobate. *Phys. Status Solidi C Conf.* **2005**, *2*, 302–305. [CrossRef]
21. Heitjans, P.; Indris, S. Fast Diffusion in Nanocrystalline Ceramics Prepared by Ball Milling. *J. Mater. Sci.* **2004**, *39*, 5091–5096. [CrossRef]
22. Heitjans, P.; Masoud, M.; Feldhoff, A.; Wilkening, M. NMR and Impedance Studies of Nanocrystalline and Amorphous Ion Conductors: Lithium Niobate as a Model System. *Faraday Discuss.* **2007**, *134*, 67–82. [CrossRef] [PubMed]
23. Bork, D.; Heitjans, P. NMR Relaxation Study of Ion Dynamics in Nanocrystalline and Polycrystalline LiNbO$_3$. *J. Phys. Chem. B* **2002**, *102*, 7303–7306. [CrossRef]
24. Subhadra, K.G.; Kishan, R.K.; Sirdeshmukh, D.B. Systematic Hardness Studies on Lithium Niobate Crystals. *Bull. Mater. Sci.* **2000**, *23*, 147–150. [CrossRef]
25. Weber, M.J. *CRC Handbook of Laser Science and Technology Supplement 2: Optical Materials*; CRC Press: Boca Raton, FL, USA, 1995; p. 18.
26. Chatterjee, M.; Naskar, M.K. Novel technique for the synthesis of lithium aluminate (LiAlO$_2$) powders from water-based sols. *J. Mater. Sci. Lett.* **2003**, *22*, 1747–1749. [CrossRef]
27. Kinoshita, K.; Sim, J.W.; Ackerman, J.P. Preparation and characterization of lithium aluminate. *Mat. Res. Bull.* **1978**, *13*, 445–455. [CrossRef]
28. Bishay, A. *Recent Advances in Science and Technology of Materials*; Plenum Press: New York, NY, USA, 1974; p. 281.
29. Bartasyte, A.; Plausinaitiene, V.; Abrutis, A.; Stanionyte, S.; Margueron, S.; Boulet, P.; Kobata, T.; Uesu, Y.; Gleize, J. Identification of LiNbO$_3$, LiNb$_3$O$_8$ and Li$_3$NbO$_4$ Phases in Thin Films Synthesized with Different Deposition Techniques by Means of XRD and Raman Spectroscopy. *J. Phys. Condens. Matter* **2013**, *25*, 205901. [CrossRef] [PubMed]
30. Brooker, M.H.; Bates, J.B. Raman and Infrared Spectral Studies of Anhydrous Li$_2$CO$_3$ and Na$_2$CO$_3$. *J. Chem. Phys.* **1971**, *54*, 4788–4796. [CrossRef]
31. Cava, S.; Tebcherani, S.M.; Souza, I.A.; Pianaro, S.A.; Paskocimas, C.A.; Longo, E.; Varela, J.A. Structural Characterization of Phase Transition of Al$_2$O$_3$ Nanopowders Obtained by Polymeric Precursor Method. *Mater. Chem. Phys.* **2007**, *103*, 394–399. [CrossRef]
32. Phillips, W.; Amodei, J.J.; Staebler, D.L. Optical and Holographic Storage Properties of Transition Metal Doped Lithium Niobate. *RCA Rev.* **1972**, *33*, 94–109.
33. Schirmer, O.F.; Imlau, M.; Merschjann, C.; Schoke, B. Electron Small Polarons and Bipolarons in LiNbO$_3$. *J. Phys. Condens. Matter* **2009**, *21*, 123201. [CrossRef]
34. Smyth, D.M. Defects and Transport in LiNbO$_3$. *Ferroelectrics* **1983**, *50*, 93–102. [CrossRef]
35. Semiletov, S.A.; Bocharova, N.G.; Rakova, E.V. Decomposition of a Solid Solution on the Surface of Lithium Niobate Crystals: Structure, Morphology, and Mutual Orientation of Phases. *Growth Cryst.* **2012**, *17*, 95–103.
36. Sugak, D.Y.; Syvorotka, I.I.; Buryy, O.A.; Yakhnevych, U.V.; Solskii, I.M.; Martynyuk, N.V.; Suhak, Y.; Suchocki, A.; Zhydachevskii, Y.; Jakiela, R. Spatial Distribution of Optical Coloration in Single Crystalline LiNbO$_3$ after High-Temperature H$_2$/Air Treatments. *Opt. Mater.* **2017**, *70*, 106–115. [CrossRef]
37. Holmes, R.J.; Minford, W.J. The effects of boule to boule compositional variations on the properties of LiNbO$_3$ electro-optic devices – An interpretation from defect chemistry studies. *Ferroelectrics* **1987**, *75*, 63–70. [CrossRef]
38. Kovács, L.; Polgár, K. Density Measurements on LiNbO$_3$ Crystals Confirming Nb Substitution for Li. *Cryst. Res. Technol.* **1986**, *21*, K101–K104. [CrossRef]
39. Lundberg, M. The Crystal Structure of LiNb$_3$O$_8$. *Acta Chem. Scand.* **1971**, *25*, 3337–3346. [CrossRef]

40. Kalinin, S.V.; Kim, Y.; Fong, D.D.; Morozovska, A.N. Surface-Screening Mechanisms in Ferroelectric Thin Films and Their Effect on Polarization Dynamics and Domain Structures. *Reports Prog. Phys.* **2018**, *81*, 36502. [CrossRef] [PubMed]
41. Jian, Z.; Lu, X.; Fang, Z.; Hu, Y.S.; Zhou, J.; Chen, W.; Chen, L. LiNb$_3$O$_8$ as a Novel Anode Material for Lithium-Ion Batteries. *Electrochem. Commun.* **2011**, *13*, 1127–1130. [CrossRef]

© 2019 by the authors. Licensee MDPI, Basel, Switzerland. This article is an open access article distributed under the terms and conditions of the Creative Commons Attribution (CC BY) license (http://creativecommons.org/licenses/by/4.0/).

Article

The Photorefractive Response of Zn and Mo Codoped LiNbO₃ in the Visible Region

Liyun Xue [1], Hongde Liu [1,*], Dahuai Zheng [2], Shahzad Saeed [1], Xuying Wang [1], Tian Tian [3], Ling Zhu [1], Yongfa Kong [1,2,*], Shiguo Liu [1], Shaolin Chen [2], Ling Zhang [2] and Jingjun Xu [1,2]

1. The MOE Key Laboratory of Weak-Light Nonlinear Photonics and School of Physics, Nankai University, Tianjin 300071, China; 1120150044@mail.nankai.edu.cn (L.X.); shehzadsaeed2003@yahoo.com (S.S.); 2120180174@mail.nankai.edu.cn (X.W.); 1120110056@mail.nankai.edu.cn (L.Z.); nkliusg@nankai.edu.cn (S.L.); jjxu@nankai.edu.cn (J.X.)
2. TEDA Institute of Applied Physics, Nankai University, Tianjin 300457, China; dhzheng@nankai.edu.cn (D.Z.); chenshaolin@nankai.edu.cn (S.C.); zhangl63@nankai.edu.cn (L.Z.)
3. Institute of Crystal Growth, School of Materials Science and Engineering, Shanghai Institute of Technology, Shanghai 201418, China; tiant@sit.edu.cn
* Correspondence: liuhd97@nankai.edu.cn (H.L.); kongyf@nankai.edu.cn (Y.K.); Tel.: +86-139-0219-7640 (H.L.); +86-138-2019-3619 (Y.K.)

Received: 2 April 2019; Accepted: 25 April 2019; Published: 28 April 2019

Abstract: We mainly investigated the effect of the valence state of photorefractive resistant elements on the photorefractive properties of codoped crystals, taking the Zn and Mo codoped LiNbO₃ (LN:Mo,Zn) crystal as an example. Especially, the response time and photorefractive sensitivity of 7.2 mol% Zn and 0.5 mol% Mo codoped with LiNbO₃ (LN:Mo,Zn$_{7.2}$) crystal are 0.65 s and 4.35 cm/J at 442 nm, respectively. The photorefractive properties of the LN:Mo,Zn crystal are similar to the Mg and Mo codoped LiNbO₃ crystal, which are better than the Zr and Mo codoped LiNbO₃ crystal. The results show that the valence state of photorefractive resistant ions is an important factor for the photorefractive properties of codoped crystals and that the LN:Mo,Zn$_{7.2}$ crystal is another potential material with fast response to holographic storage.

Keywords: lithium niobate; optical storage materials; photorefractive materials

1. Introduction

Volume holographic storage is a technology that can store information at high density inside a crystal, which exploits the "inside" of the storage material, rather than only using its surface, resulting in a massive increase in the capacity of the data storage. Lithium niobate (LN) crystal is an essential material in many applications [1–5], such as surface acoustic wave, waveguides, volume holographic storage, piezoelectric, pyroelectric, and integrated optics [6–9]. The performance of volume holographic storage is particularly prominent [10–18]. However, there are still some problems in applying it to real life. One of the critical issues is that the holographic response of existing bulk holographic materials is not fast enough. The photorefractive performance of the LN crystal can be improved by using the doping technique.

In 2012, the photorefraction of Mo-doped LN crystals was reported. T. Tian et al. found that Mo⁶⁺ doping promotes the photorefractive properties of LN crystals [19]. Then in 2013, they studied fast UV-Vis photorefractive response of Zr and Mg codoped LiNbO₃:Mo and found that their experimental phenomena were very different [20]. To explore the effect of the valence state of optical-damage-resistant ions on the properties of codoped crystals, in 2018, L. Zhu et al. investigated into In and Hf codoped LiNbO₃:Mo crystals [21,22]. They proposed that these crystals, likewise, can improve the photorefractive properties of codoped LN crystals. Although Zr and Mg, In, Hf are elements with optical damage resistance in LN crystal, their effects on the photorefraction of LN:Mo crystal are different. To further study the effect of the optical damage resistant on the photorefractive properties of the codoped LN crystals, and explore the specific defect structure inside the crystal, we chose Zn and Mo codoped LN crystals.

2. Materials and Methods

We grew crystals by using the Czochralski method. The raw materials used were Li_2CO_3, Nb_2O_5, MoO_3, and ZnO with a purity of 99.99%. The Li/Nb ratio in the initial melt was 48.38/51.62. According to the previous experimental results, it was found that 0.5 mol% Mo was superior in the monodoped lithium niobate crystal, and the doping threshold of Zn was 7 mol%. So, the concentration we chose of MoO_3 was 0.5 mol%, and the concentrations of ZnO were 5.4 and 7.2 mol%, respectively. For comparison, the congruent lithium niobate (CLN) and monodoped with Mo lithium niobate crystals were also grown. The crystals were labeled as CLN, LN:$Mo_{0.5}$, LN:Mo,$Zn_{5.4}$, and LN:Mo,$Zn_{7.2}$. Before preparing polycrystals, these materials were thoroughly mixed and sintered. The conditions for crystal growth were a pulling rate of 0.6 mm/h and a rotation speed of 14 r/min. The boules were about 4.0 cm long and had a diameter of about 3.5 cm. The subsequent polarization was applied in the furnace, with a polarization current of 30 mA for 15 min. The crystals were then cut into 3 mm and 1 mm-thick plates after the annealing treatment, and optically polished in the y-face for optical measurement.

We measured the diffraction efficiency and response time of the crystals by the two wave coupling holographic experiment as shown in Figure 1. Two writing beams of extraordinary light with equal light intensity were employed to write the holograms, and their crossing angle was designed to be 30° in air. The instruments used in the experiments were helium cadmium laser (442 nm), argon ion laser (488 nm), and semiconductor solid state laser (532 nm and 671 nm). The total light intensities used here were 250 mW/cm^2 for 442 nm, 400 mW/cm^2 for 488 nm and 532 nm, and 1200 mW/cm^2 for 671 nm. The measured diffraction efficiency was defined as $\eta = I_d/(I_d + I_t)$, where I_d and I_t is the diffracted and transmitted light intensities of the readout beam, respectively. The response time constant τ_r and the saturation diffraction efficiency η_s were deduced by fitting the function $\eta_{(t)} = \eta_s[1 - \exp(-t/\tau_r)]^2$ to the data. The photorefractive sensitivity [23] was defined as $S = (d\sqrt{\eta}/dt)_{t=0}/(IL)$, where I is the recording intensity and L is the grating thickness. The UV-visible transmission spectrum of the crystals was measured using a UV-4100 spectrophotometer (Hitachi Science and Technology, Tokyo, Japan) with a range of 300–800 nm and a resolution of 1 nm/step. The measurement of the infrared spectrum was carried out by A MAGNA-560 FT-IR spectrometer (Thermo Nicolet Corporation, Madison, America) with a range of 400–4000 cm^{-1} and a resolution of 2 cm^{-1}/step.

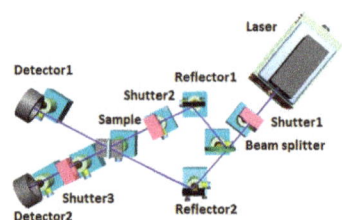

Figure 1. The schematic diagram of the experimental two-wave holographic setup.

3. Results

3.1. Photorefractive Properties

Figure 2a–d show the typical curves of diffraction efficiencies for the LN:Mo,$Zn_{7.2}$ crystal as a function of time at 442 nm, 488 nm, 532 nm, and 671 nm lasers, respectively. The saturation diffraction efficiencies of the LN:Mo,$Zn_{7.2}$ crystal to each wavelength were 4.50%, 17.72%, 2.32%, and 0.19%, respectively. From the figures, we can see that the diffraction efficiencies of the LN:Mo,$Zn_{7.2}$ crystal increased with the shortening of the wavelength, but the mutation occurred at 488 nm, where the diffraction efficiency is four times that of 442 nm. The reason for this is probably due to the lower power of the 442 nm laser, which was the maximum laser power in our lab. However, the crystal had different responsiveness to each band. Among the lasers used in our experiments, especially in the 671 nm band, the required intensity was the highest.

The response times of the LN:Mo,Zn$_{7.2}$ crystal were 0.65 s, 3.57 s, 1.71 s, and 6.61 s under the action of 442 nm, 488 nm, 532 nm, and 671 nm lasers, respectively. As the laser wavelength decreased, the response time of the LN:Mo,Zn$_{7.2}$ crystal gradually decreased, but the response time at 488 nm was about two times longer than the response time taken at 532 nm. The reason for this phenomenon is likely to be related to its higher diffraction efficiency, so we used photorefractive sensitivity to reflect the comprehensive photorefractive ability of the LN:Mo,Zn$_{7.2}$ crystal.

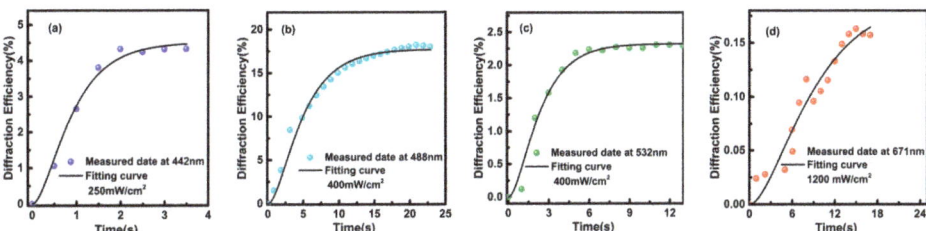

Figure 2. The diffraction efficiency of the 7.2 mol% Zn and 0.5 mol% Mo codoped with LiNbO$_3$ (LN:Mo,Zn$_{7.2}$) crystal versus time at (**a**) 442 nm, (**b**) 488 nm, (**c**) 532 nm, and (**d**) 671 nm laser, respectively.

Figure 3 shows the photorefractive properties of LN:Mo,Zn crystals at 442 nm, 488 nm, 532 nm, and 671 nm laser wavelengths. For comparison, experimental data of CLN crystals is also included. The saturation diffraction efficiency is depicted in Figure 3a, the response time in Figure 3b, and the photorefractive sensitivity in Figure 3c. It can be seen from Figure 3a that the saturation diffraction efficiency of the LN:Mo,Zn crystal was improved due to the addition of zinc. In particular, LN:Mo, Zn$_{7.2}$ had better performance in each band than other crystals and was more special at 488 nm. LN:Mo$_{0.5}$ crystal had the highest saturation diffraction efficiency. Moreover, for all doped samples, the saturation diffraction efficiencies at 488 nm were higher than that at 442 nm. As shown in Figure 3b, the response time of the codoped crystal was greatly shortened due to the incorporation of Zn. In order to obtain a comparison of the crystals synthesis effects, we used the photorefractive sensitivity of the crystals to analyze that. Figure 3c shows that the photorefractive sensitivity of all our crystals increased with the shortening of the wavelength. When the zinc element exceeded the threshold, the photorefractive sensitivity of the LN:Mo,Zn crystal in different bands was improved. The photorefractive sensitivity of the LN:Mo,Zn$_{7.2}$ crystal was significantly better than other crystals in each wavelength band. The photorefractive sensitivity of the LN:Mo,Zn$_{7.2}$ crystal was 4.35 cm/J, 0.98 cm/J, 0.74 cm/J, and 0.02 cm/J at 442 nm, 488 nm, 532 nm, and 671 nm lasers, respectively. Compared with the data of LN:Mo,Zr$_{2.5}$ and LN:Mo,Mg$_{6.5}$ in Reference [20] in the same wavelength, the zinc doping enhanced the photorefractive properties of LN:Mo crystal, and its variation phenomena was the same as that of LN:Mo,Mg crystals, and was opposite to LN:Mo,Zr crystals. This indicates that the valence state of optical damage resistant ions plays a key role.

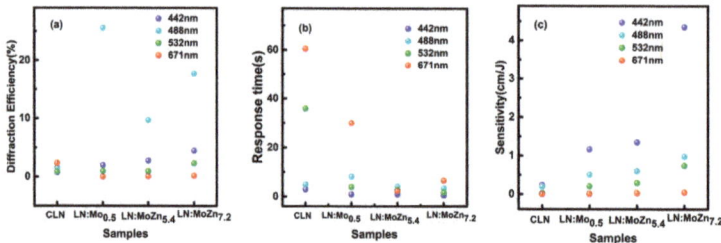

Figure 3. (**a**) The diffraction efficiency, (**b**) response time, and (**c**) photorefractive sensitivity of CLN, LN:Mo$_{0.5}$, LN:Mo,Zn$_{5.4}$, and LN:Mo,Zn$_{7.2}$ crystals at 442 nm, 488 nm, 532 nm, and 671 nm lasers, respectively.

3.2. Spectral Analysis

The spectral characteristics of the crystals in different wavelengths can reveal the defect structure of the crystals. We recorded the wavelength of the absorption coefficient at 20 cm^{-1} as the absorption edge of the crystal. From the inset of Figure 4a, the absorption edge of the LN:Mo,Zn$_{5.4}$ crystal had a blue-shift owing to the ZnO doping. The LN:Mo,Zn$_{7.2}$ crystal occurred a red-shift for its doping concentration exceeding the threshold, but this move was very tiny. It can be seen from the UV-Vis absorption difference spectrum of Figure 4a, which indicates the difference in absorption between the test crystals and the CLN crystal, that the LN:Mo,Zn crystals have an absorption valley near 319 nm, and their values were very close to the absorption edge of the LN:Mo,Zn crystal. According to X. Li et al. research, the absorption edge of pure LN crystal is caused by the defect of lithium vacancy (V_{Li}^-) [24]. Therefore, we speculate that the absorption valley was caused by V_{Li}^-. There is a distinct absorption peak near 330 nm and 480 nm, which is similar to the LN:Mo,Mg and LN:Mo,Zr crystals. This is most likely due to the deep impurity level introduced by Mo ions. According to the existing energy band information of the ABO$_3$ ferroelectric crystal, we know that for LN crystals, the valence band top is formed by the 2p orbital of oxygen, and the lowest layer of the empty conduction band is provided by the 4d orbit of the transition metal Nb^{5+} ions. Molybdenum ions and niobium ions belong to transition metal ions and have a 4d orbital. When molybdenum ions are incorporated into the LN crystal, its orbits will affect the bonding strength of the Nb-O bond, resulting in a change in the band gap width of the LN crystal. The absorption band of the visible band indicates that there are multiple energy levels in the band gap which contribute to the photorefractive processing.

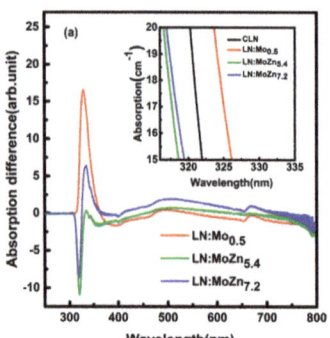
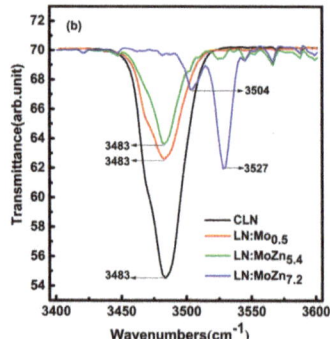

Figure 4. (**a**) UV-visible absorption difference spectra of LN:Mo, LN:Mo,Zn relative to CLN crystals, and the inset shows absorption edges at 20 cm^{-1}. (**b**) FT-IR transmission spectra of CLN, LN:Mo and LN:Mo,Zn crystals.

From the FT-IR transmission spectrum which is shown in Figure 4b, we can observe that in the LN:Mo,Zn crystals, the main absorption peak of LN:Mo,Zn$_{7.2}$ crystal appeared at 3527 cm^{-1}, the minor peak appeared at 3504 cm^{-1}, and the peak appeared at 3483 cm^{-1} in CLN, LN:Mo$_{0.5}$ and LN:Mo,Zn$_{5.4}$. The absorption peak transferred from 3483 cm^{-1} to 3527 cm^{-1}, which implies that the Zn^{2+} concentration in LN:Mo,Zn$_{7.2}$ crystal exceeded the threshold, the (Nb$_{Li}$)$^{4+}$ defects disappeared, and the (Zn$_{Nb}$)$^{3-}$ appeared in the crystal.

To detect the defect structures inside the crystal, we analyzed the valence of Mo ions of LN:Mo,Zn$_{7.2}$ crystal by the X-ray photoelectron spectroscopy (XPS) [25]. As shown in Figure 5a, molybdenum ions exhibited three valence states in the LN:Mo,Zn$_{7.2}$ crystal, in which the combined peaks of 233.5 eV and 236.7 eV represent +6 valence, the peaks of 231 eV and 234.5 eV represent +5 valence, and the peaks of 229.3 eV and 232.5 eV represent +4 valence. As a comparison, the molybdenum element in the residue of the crucible only showed a +6 valence state formed by two combined peaks of 232.6 eV and 235.7 eV shown in Figure 5b. These results show that the valence states of Mo ions changed when they enter the crystals, which lead to the generation of new defects.

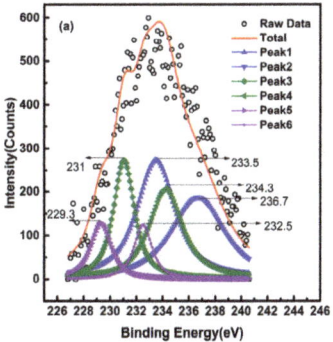 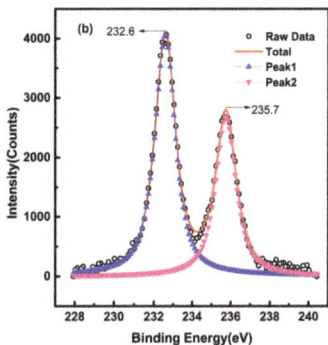

Figure 5. X-ray photoelectron spectroscopy of Mo in LN:Mo,Zn$_{7.2}$ crystal (**a**) in the crystal and (**b**) in the residue of the crucible.

3.3. Optical Damage Resistance Ability

We used the light spot distortion method to measure the optical-damage-resistance ability of crystals. As shown in Figure 6, at the maximum laser power in our lab, we can observe that the transmitted spots of the LN:Mo,Zn$_{7.2}$ crystal remained circular, under the effect of 671 nm laser, when the light intensity was maintained at an intensity of 7.8 × 10^4 W/cm^2 for 5 min. A similar phenomenon occurred when under the action of the 532 nm and 488 nm lasers, the light intensities were 2.6 × 10^5 W/cm^2 and 3.2 × 10^5 W/cm^2, respectively. However, the transmitted spot of the LN:Mo,Zn$_{5.4}$ crystal was significantly stretched in the c-axis direction compared to the original incident one. This performance broadens the range of applications for the LN:Mo,Zn$_{7.2}$ crystal, allowing it to operate at higher light levels. The above experiment shows that the LN:Mo,Zn$_{7.2}$ crystal not only has realized itself as an excellent photorefractive material, but also a high-intensity-application crystal.

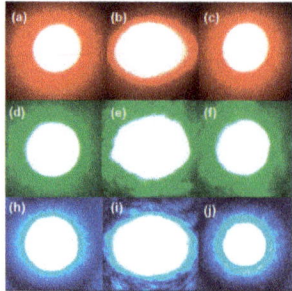

Figure 6. The incident spots of (**a**) 671nm, (**d**) 532nm, and (**h**) 488nm lasers; (**b,e,i**) transmitted spots of LN:Mo,Zn$_{5.4}$ crystals under the action of different color lasers; (**c,f,j**) transmitted spots of LN:Mo,Zn$_{7.2}$ crystal under the action of different color lasers.

4. Discussion

In the absorption difference spectrum of Figure 4a, we can observe that the absorption near 488 nm is higher than other spectra in the visible band, which also confirms that it has higher diffraction efficiency than other bands. Also, the apparent absorption peak near 330 nm indicates that there may be a deep defect level here. Because of the low power of our laboratory 325 nm laser and the absence of working laser of suitable wavelength, further research on deep defect level is needed.

The XPS studies reveal that the valence change of Mo ions occurs before and after entering the LN:Mo,Zn crystal. In the process of crystal growth, the convertible Mo ions enter the crystal and are isolated from oxygen, so they are not easily oxidized and present three valence states. On the contrary, the Mo ions in the residue can come in contact with sufficient oxygen and be easily oxidized, so the Mo ions in the residue only exist in the +6 valence state. According to previous reports [17], the Mo^{4+} and Mo^{5+} ions may occupy the Li sites (Mo$_{Li}^{3+/4+}$), and Mo^{6+} ions may occupy the Nb sites (Mo$_{Nb}^{+}$) severed as UV photorefractive centers. On the base of the lithium vacancy model, a large number of intrinsic defects (such as Nb$_{Li}^{4+}$, small polaron, and bipolaron, etc.) limit the response time of the crystal. At low concentrations, the molybdenum and zinc ions tend to occupy the Li position, which could push the Nb$_{Li}^{4+}$ out and shorten the response time. When the zinc ion concentration exceeds the threshold, all the Nb$_{Li}^{4+}$ were replaced and the Mo$_{Li}^{3+/4+}$ were also substituted due to their higher valence than +2 valence zinc ions and transferred to the Mo$_{Nb}^{+}$ ions. As the UV photorefractive center, the Mo$_{Nb}^{+}$ ions may attract an electron from O^{2-}, which could further speed up the photorefraction process [26]. However, for the +3 and +4 valence photorefractive resistant elements, the substitution for the Mo$_{Li}^{3+/4+}$ defect is weak. Thus, the valence state of the optical damage resistant elements may be the critical factor determining the crystal properties.

As the main obstacle of the optical storage materials, the storage speed has been limiting the commercial application of LN crystals. The response time and sensitivity of LN:Mo,Zn crystal was optimized by the zinc codoping, compared with LN:Mo,Zr crystals, which is similar to the LN:Mo,Mg crystals. These results confirm that the valence state of the optical damage resistant elements may be the critical factor determining the crystal properties. Compared to the LN:Mo,Mg$_{6.5}$ crystals, the response time in LN:Mo,Zn$_{7.2}$ crystals was several times longer than that of LN:Mo,Mg$_{6.5}$ crystals at the same wavelength. Especially with the shortening of wavelength, this gap was further widened. We think this due to the individuality of the elements, such as the ionic radius, electronegativity, outer electron configuration, etc. As it is well known, the conduction band is generally provided by d electrons. For Zn^{2+} and Mg^{2+}, Zn^{2+} has d electrons, while Mg ions have no d electrons, which results in a significant difference in the effect of the two on the crystal. Overall, the valence state of the optical-damage-resistant elements and the individuality of the elements may be the main factor determining the crystal properties.

5. Conclusions

In summary, the LN:Mo,Zn crystals were grown with different concentrations and measured about photorefractive properties. Through the experiments, we find that its photorefractive properties are similar to those of LN:Mo,Mg crystals, but different from those of LN:Mo,Zr crystals, and the doping of zinc can shorten the response time and improve the photorefractive sensitivity of the LN:Mo,Zn crystal, especially 0.65 s and 4.35 cm/J at 442nm for the LN:Mo,Zn$_{7.2}$ crystal. It is indicated that the valence state of ions has a significant effect on the photorefractive properties of crystals. The valence states of Mo ions in crystal was identified by the XPS results, and the Mo$_{Nb}^+$ and Mo$_{Li}^{3+/4+}$ defects were served as the photorefractive center for fast photorefraction. The experimental results show that the LN:Mo,Zn$_{7.2}$ crystal can be used as another candidate material for holographic storage.

Author Contributions: Conceptualization, L.X., H.L. and Y.K.; Funding acquisition, D.Z., T.T. and Y.K.; Investigation, L.X.; Methodology, L.X., D.Z., S.L., S.C., and L.Z.; Project administration, Y.K. and J.X.; Resources, L.X., S.L., S.C. and J.X.; Writing (original draft), L.X.; Writing (review and editing), L.X., H.L., D.Z., S.S., X.W., L.Z., and Y.K.

Funding: This research was partially supported by the National Natural Science Foundation of China (11674179, 61705116, and 61605116) and the Program for Changjiang Scholars and Innovative Research Team in University with grant [IRT_13R29], National Basic Research Program of China (2013CB328706), and International Science & Technology Cooperation Program of China (2013DFG52660).

Conflicts of Interest: The authors declare no conflicts of interest.

References

1. Wang, C.; Zhang, M.; Stern, B.; Lipson, M.; Lončar, M. Nanophotonic lithium niobate electro-optic modulators. *Opt. Express* **2018**, *26*, 1547–1555. [CrossRef] [PubMed]
2. Krasnokutska, I.; Tambasco, J.J.; Li, X.; Peruzzo, A. Ultra-low loss photonic circuits in lithium niobate on insulator. *Opt. Express* **2018**, *26*, 897–904. [CrossRef] [PubMed]
3. Wang, C.; Li, Z.; Kim, M.H.; Xiong, X.; Ren, X.F.; Guo, G.C.; Yu, N.; Lončar, M. Metasurface-assisted phase-matching-free second harmonic generation in lithium niobate waveguides. *Nat. Commun.* **2017**, *8*, 4–10. [CrossRef] [PubMed]
4. Hesselink, L.; Orlov, S.S.; Liu, A.; Akella, A.; Lande, D.; Neurgaonkar, R.R. Photorefractive materials for nonvolatile volume holographic data storage. *Science* **1998**, *282*, 1089–1094. [CrossRef]
5. Xu, M.; Kang, H.; Guan, L.; Li, H.; Zhang, M. Facile Fabrication of a Flexible LiNbO$_3$ Piezoelectric Sensor through Hot Pressing for Biomechanical Monitoring. *ACS Appl. Mater. Interfaces* **2017**, *9*, 34687–34695. [CrossRef] [PubMed]
6. Qiu, W.; Ndao, A.; Vila, V.C.; Salut, R.; Courjal, N.; Baida, F.I.; Bernal, M.P. Fano resonance-based highly sensitive, compact temperature sensor on thin film lithium niobate. *Opt. Lett.* **2016**, *41*, 1106–1109. [CrossRef] [PubMed]
7. Xu, H.; Dong, S.; Xuan, W.; Farooq, U.; Huang, S.; Li, M.; Wu, T.; Jin, H.; Wang, X.; Luo, J. Flexible surface acoustic wave strain sensor based on single crystalline LiNbO$_3$ thin film. *Appl. Phys. Lett.* **2018**, *112*, 093502. [CrossRef]
8. Bai, Y.; Kachru, R. Nonvolatile holographic storage with two-step recording in lithium niobate using cw lasers. *Phys. Rev. Lett.* **1997**, *78*, 2944. [CrossRef]
9. Yoshinaga, H.; Kitayama, K.i.; Oguri, H. Holographic image storage in iron-doped lithium niobate fibers. *Appl. Phys. Lett.* **1990**, *56*, 1728–1730. [CrossRef]
10. Imlau, M.; Brüning, H.; Schoke, B.; Hardt, R.; Conradi, D.; Merschjann, C. Hologram recording via spatial density modulation of Nb$_{Li}^{4+/5+}$ antisites in lithium niobate. *Opt. Express* **2011**, *19*, 15322–15338. [CrossRef]
11. Juodkazis, S.; Mizeikis, V.; Sūdžius, M.; Misawa, H.; Kitamura, K.; Takekawa, S.; Gamaly, E.G.; Krolikowski, W.Z.; Rode, A.V. Laser induced memory bits in photorefractive LiNbO$_3$ and LiTaO$_3$. *Appl. Phys. A* **2008**, *93*, 129–133. [CrossRef]
12. Camarillo, E.; Murrieta, H.; Hernandez, J.M.; Zoilo, R.; Flores, M.C.; Han, T.P.J.; Jaque, F. Optical properties of LiNbO$_3$:Cr crystals co-doped with germanium oxide. *J. Lumin.* **2008**, *128*, 747–750. [CrossRef]

13. Luo, S.; Meng, Q.; Wang, J.; Sun, X. Effect of In^{3+} concentration on the photorefraction and scattering properties in In:Fe:C$_U$:LiNbO$_3$ crystals at 532nm wavelength. *Opt. Commun.* **2016**, *358*, 198–201. [CrossRef]
14. Nie, Y.; Wang, R.; Wang, B. Growth and holographic storage properties of In:Ce:Cu:LiNbO$_3$ crystal. *Mater. Chem. Phys.* **2007**, *102*, 281–283. [CrossRef]
15. Zhen, X.H.; Li, H.T.; Sun, Z.J.; Ye, S.J.; Zhao, L.C.; Xu, Y.H. Holographic properties of double-doped Zn:Fe:LiNbO$_3$ crystals. *Mater. Lett.* **2004**, *58*, 1000–1002. [CrossRef]
16. Wei, Z.; Naidong, Z.; Qingquan, L. Growth and Holographic Storage Properties of Sc, Fe Co-Doped Lithium Niobate Crystals. *J. Rare Earth.* **2007**, *25*, 775–778. [CrossRef]
17. Xu, C.; Leng, X.; Xu, L.; Wen, A.; Xu, Y. Enhanced nonvolatile holographic properties in Zn, Ru and Fe co-doped LiNbO$_3$ crystals. *Opt. Commun.* **2012**, *285*, 3868–3871. [CrossRef]
18. Pálfalvi, L.; Hebling, J.; Almási, G.; Péter, Á.; Polgár, K.; Lengyel, K.; Szipöcs, R. Nonlinear refraction and absorption of Mg doped stoichiometric and congruent LiNbO$_3$. *J. Appl. Phys.* **2004**, *95*, 902. [CrossRef]
19. Tian, T.; Kong, Y.; Liu, S.; Li, W.; Wu, L.; Chen, S.; Xu, J. Photorefraction of molybdenum-doped lithium niobate crystals. *Opt. Lett.* **2012**, *37*, 2679. [CrossRef]
20. Tian, T.; Kong, Y.; Liu, S.; Li, W.; Chen, S.; Rupp, R.; Xu, J. Fast UV-Vis photorefractive response of Zr and Mg co-doped LiNbO$_3$:Mo. *Opt. Express* **2013**, *21*, 128–132. [CrossRef]
21. Zhu, L.; Zheng, D.; Saeed, S.; Wang, S.; Liu, H.; Kong, Y.; Liu, S.; Chen, S.; Zhang, L.; Xu, J. Photorefractive Properties of Molybdenum and Hafnium Co-Doped LiNbO$_3$ Crystals. *Crystals* **2018**, *8*, 322. [CrossRef]
22. Zhu, L.; Zheng, D.; Liu, H.; Saeed, S.; Wang, S.; Liu, S.; Chen, S.; Kong, Y.; Xu, J. Enhanced photorefractive properties of indium co-doped LiNbO$_3$:Mo crystals. *AIP Adv.* **2018**, *8*, 095316. [CrossRef]
23. Dischler, B.; Herrington, J.R.; Räuber, A.; Kurz, H. Correlation of the photorefractive sensitivity in doped LiNbO$_3$ with chemically induced changes in the optical absorption spectra. *Solid State Commun.* **1974**, *14*, 1233–1236. [CrossRef]
24. Li, X.; Kong, Y.; Liu, H.; Sun, L.; Xu, J.; Chen, S.; Zhang, L.; Huang, Z.; Liu, S.; Zhang, G. Origin of the generally defined absorption edge of non-stoichiometric lithium niobate crystals. *Solid State Commun.* **2007**, *141*, 113–116. [CrossRef]
25. Grim, S.O.; Matienzo, L.J. X-Ray Photoelectron Spectroscopy of Inorganic and Organometallic Compounds of Molybdenum. *Inorg. Chem.* **1975**, *14*, 1014–1018. [CrossRef]
26. Kong, Y.; Xu, J.; Zhang, W.; Zhang, G. The site occupation of protons in lithium niobate crystals. *J. Phys. Chem. Solids* **2000**, *61*, 1331. [CrossRef]

© 2019 by the authors. Licensee MDPI, Basel, Switzerland. This article is an open access article distributed under the terms and conditions of the Creative Commons Attribution (CC BY) license (http://creativecommons.org/licenses/by/4.0/).

MDPI
St. Alban-Anlage 66
4052 Basel
Switzerland
Tel. +41 61 683 77 34
Fax +41 61 302 89 18
www.mdpi.com

Crystals Editorial Office
E-mail: crystals@mdpi.com
www.mdpi.com/journal/crystals

www.ingramcontent.com/pod-product-compliance
Lightning Source LLC
LaVergne TN
LVHW070551100526
838202LV00012B/437